GROUP PRACTICE IN DESIGN

By the same author: EDUARDO PAOLOZZI

MICHAEL MIDDLETON

GROUP PRACTICE IN DESIGN

With a Note by MISHA BLACK,
OBE, RDI, PPSIA, MInst RA

THE ARCHITECTURAL PRESS, LONDON

85139 260 1

First published 1967

Made and printed in Great Britain by
William Clowes and Sons, Limited, London and Beccles

Contents

Illustrations

Preface

This is a book about how things get designed. It pretends neither to scholarship nor to the expertise of the text-book. It is an attempt to consider the design process in a relatively broad context and to indicate the complexities of organization and communication that must nowadays be overcome if our environment is not to disintegrate completely. Historians and practising designers will find no new thoughts here. I hope, however, that some of those less closely involved may find that the book illuminates a little a process that is largely hidden from view and which, as a result, is too readily taken for granted.

I am grateful to the many people who have helped me knowingly and unknowingly. As will be clear, I have pillaged extensively—and inevitably—from Dr Nikolaus Pevsner and Dr Reyner Banham, whose *Pioneers of the Modern Movement* and *Theory and Design in the First Machine Age* respectively are basic documents for students of design. My thanks are due to the editor of the *Journal of the Royal Society of Arts* for permission to quote from Sir Gerald Barry's 1952 Cantor lectures on the Festival of Britain; the proprietors of *Time/Life* for permission to make use of material on Skidmore, Owings and Merrill which appeared in *Fortune* magazine in 1958; to the editors of *Program*, Columbia University, for permission to use extracts of an address by Dr Walter Gropius; and to the editors of the *Architectural Review*, and all others acknowledged in the text, to quote from their pages.

Among the many individuals who generously gave valuable time to provide detailed information and, when pestered, smilingly produced more, I must gratefully acknowledge in particular: Mr Henry Dreyfuss and his partner Mr William Purcell; Mr Michael Powell, Mr A. H. Withers and Mr R. W. Robson-Smith, all of the Architect's Department of the Greater London Council; Sig Alberto Rosselli of the Ponti/Fornaroli/Rosselli studio; Mr Peter Jackson of P & O/Orient line and Mr Milner Gray, Mr Kenneth Bayes and Miss Dorothy Goslett of Design Research Unit; Mr Richard Levin, Mr Clifford Hatts and Mr Tony Abbott, all of BBC Television's Design Department, the last two of whom were also good enough to provide me with photographs of their own taking; Mr David Rock, of Building Design Partnership. I have attempted to acknowledge lesser textual debts in the body of the book; if any has been overlooked may I offer to its author my sincere apologies. Many sources have been helpful and generous with illustration material. Mr Peter Kneebone has provided from a private *cache* material relating to the Festival of Britain which, I suspect, is now available from no other source; Mr F. H. K. Henrion kindly made available a number of his own photographs of the Countryside pavilion. My grateful thanks are due to Mr Frederick Dickinson and Mr Cluny Gillies for the skill and perseverance with which they located many of the illustrations. Thanks are due also to all who have given permission to reproduce illustrations: they are named on pages 299–300.

The book owes its origin to Professor Misha Black who, with his usual disarming skill, talked me into it against my better judgement longer ago than I care to recall. If the result bears only a limited resemblance to what we first discussed, its deficiencies will in part be offset by Professor Black's own bracing contribution on the realities of group existence. For this, and his support throughout, I owe him my warmest thanks.

MICHAEL MIDDLETON

1. The Challenge of the Past

Who designed the columns of the Parthenon; who Chartres? By what collaborative magic sprang Venice from the waters in all her unified diversity, long centuries before our cumbersome machinery of by-laws and elevational control? Names come to us dimly, filtered by time—some as immortal as the gods they served; some rediscovered; some invented, it may be, in the sense of Voltaire's Deity, by historians. Ancient records detail their expenditure on now forgotten buildings. Their status is enshrined in the Charters of craft guilds. But of their system of working, of their exact creative relationship one with another, of their studio and workshop practices, we know next to nothing. All is conjecture. And yet the unity of artistic expression achieved by these teams of often unknown craftsmen remains to mock us; to mock and inspire us with a romantic dream of an ideal unity which sometimes seems within our grasp, yet always just eludes us; perhaps always will.

It is inevitable that our view of achievements long past is distorted by the historical perspective down which we look. Distances in time become compressed, like the distances between objects seen through a telephoto lens. The changes in the climate of taste which took place between the mid-twelfth century and the first quarter of the thirteenth, as reflected in the illustrations to the Winchester Bible, are less apparent to the modern eye, we may assume, than they were to the contemporary. Examples abound of oddities, insensitivities and blundering arrogances of design which would cause us acute unease were it not for familiarity and the patina of time. How many of those who throng the Riva dei Schiavoni notice the curious lack of alignment in the windows of the southern elevation of the Doges' Palace? Does Philander's sixteenth-century capping of the Gothic west front of the cathedral at Rodez with a complete miniature Roman church front seem to us quite as preposterous as it would were we to focus upon it the same direct judgement we apply to buildings of our own time? Mozart found Nuremberg ugly. Walpole was sickened by 'Mr Adam's gingerbread and sippets of embroidery'. Molière wrote with contempt of the

> . . . fade goût des monuments gothiques,
> Ces monstres odieux des siècles ignorants.

In our day historicism intervenes. It is as though the results of the processes of history were seen at a remove, through a gauze, as beyond recall. Old buildings are accepted as part of the scenery, like a hill or a lake. They have been lent a spurious air of inevitability merely by the fact of their age.

So it is, too, with the centres of our historic cities, wherein the sentimentalist would so often call a halt, if he could, to the passage of time and would freeze them as they are, for ever. It is as unreal to hope that more than a handful of such areas, deprived of change in a changed society, could retain their life as it is to see in their heterogeneous building types and styles a consciously contrived unity. The concept of preservation is relatively recent. Previous generations were assailed by no debilitating lack of faith in their own powers; they pulled down and rebuilt as necessary, in the manner of their day. It is only the historical perspective which so often lends to the result an air of pre-ordained unity.

The picture of an essential harmony informing the creative efforts of other days is not, that is to say, as universally accurate as is sometimes loosely assumed. Even during periods when the creative spirit was riding high, the taste of the ancients was by no means infallible: the gallimaufry of sculptured fights, nymphs, boar hunts, sacred couches, infant Zeuses, local amazons, tusks, hides, fetters and shields with which the much-admired Temple of Athena Alea at Tegea was decorated, must have been bizarre, one feels, by any objective standard. Nonetheless, when all allowances have been made for this romanticizing distortion induced by the historical perspective, it remains clear that such periods have existed: periods in which different skills and disciplines have worked together within a framework of commonly held values, beliefs, rules, traditions, to produce jointly an end result of extraordinary harmony. Most of the greatest works of man have stemmed from such periods and today, as our society grows ever more specialized, fragmented and disorientated, the problem of re-establishing a measure of such unity throughout our total environment assumes paramount importance. When Ulysses came home from Troy he exclaimed that, having seen the cities of men, he knew their minds. Man-made things indeed reveal the man. It is surely no less true that the environment he has shaped for himself in turn shapes his children. 'Affected by all the objects that surround them', wrote Plutarch, 'our senses cannot avoid being influenced either for good or for ill.' It is not pleasant to consider the effect of the world we are creating upon the outlook of future generations.

The ancient world provided, in Egypt, Greece and Rome, examples of this homogeneity of creative endeavour. The spiritual revolutions at the end of the fifth and beginning of the thirteenth centuries produced the masterpieces of Byzantium and the Gothic cathedrals. The Baroque and Rococo periods attained an identity of design thinking far beyond our reach. The France of le Roi Soleil achieved it by regimentation; the England of the Georgian squares and the Regency crescents achieved it by instinct and common sense.

Of what is it that we speak, this aesthetic concord so complete that its expression appears inevitable and final? Alberti's definition of beauty will still do duty: it is 'a harmony of all the parts, in whatsoever subject it appears, fitted together with such proportion and connection that nothing could be added, diminished or altered, but for the worse'. It is a total concept, every element of which would seem to have been forged in the same moment of creative heat. It is concerned with joint relationships; with matters of scale and proportion; with the spring and radius of curves in juxtaposition, the balance of mass and void, the marriage of structure and decoration; with iconography and appropriateness and questions of material, colour and texture. It is the correspondence which leaps the open space between the fluted columns of the Parthenon and the draperies of the caryatids of the Erechtheum; it is the framework of mathematical proportion that controls every aspect of Greek art and architecture. It is the identity of aesthetic intention which makes the capitals of S. Vitale in Ravenna, with their lacy stonework grills caging a glimmering inner layer of darkness, so precise a counterpart in ornament to the spatial effect of the central arcaded apses ringing the core of the church against the outer layer of the ambulatory into which they open. It is the fusion of thought which produced the great encrusted temples of India, for example the south gopuram at Madura, wherein it is impossible to conceive of the architecture separately from the seething mass of deep carved reliefs which smothers it or, as in the internal piers, which forms it. It is St Teresa in the Cornaro Chapel, Santa Maria della Vittoria, set apart in visionary isolation from the watching cardinals, mysteriously illuminated by the gilded timber rays bringing down a shaft of daylight from an unseen source; the plaster clouds and angels in the painted heavens above, parting to suffuse the vault of the chapel with glory: an interpenetration of architecture and sculpture, lighting and decoration, such that they cease to have an independent existence and coalesce into a single exploration of revelation and reality. It is the quality which reached its most perfect expression in the Gothic cathedrals: the total dissolution of barriers between spirit and function, architecture and engineering, plan and detail. It is the affinity between the general character of a Gothic arch, its profile and the outline of its moulding; between ground plan and section, solid and void, light and shade, sculpture and glass. Examples can be adduced from many ages, from many countries. They represent the supreme creative synthesis of which man collectively is capable.

Before such glories we stand in awe, for between the aspirations and assumptions which gave them birth and those of our own time lies a gulf that is not to be bridged. No man will ever again know what it meant to be a priest in the Fifth Dynasty, or a citizen in Athens, or even to design masques for the court of King James. The great collaborative masterpieces of the past were only made possible by the existence of a common authority which gave stability to the endeavours of the creative artist—sets of references either accepted voluntarily (as in Greece, or again, say, in the thirteenth and fourteenth centuries) or else imposed by an autocracy (as in ancient Egypt or in France under Louis XIV). Not that a hard distinction can be made. Only the hack craftsman can accept a wholly dictated solution to his problem; only genius can wholly disregard the traditions into which he is born. The broad stream of creative activity at any given moment, that is to say, is guided by rules and conventions which are, in part, accepted willingly by the artist; in part, imposed upon him by social or other circumstances—if not by a political élite then more than probably, to a degree of which he may well himself be unconscious, by a cultural élite.

Such conventionalized systems of thought have sometimes been long lived. In ancient Egypt design formulae hardened into an inflexible tradition which outlasted twenty dynasties. The language of classical architecture survived for more than two millennia as a mode of expression used and understood in all parts of the civilized world: the 'most comprehensive and stable manner of design the world has ever seen'.* Disciplines of this kind exert their own co-ordination and control. At least three architects of consequence—Palladio, Buora and Longhena—worked on the church and monastery of S. Giorgio Maggiore and, by reason of the discipline of the Orders, to magically harmonious effect. Without such a common language of appearances, collaboration must end in anarchy.

In any wide sense, the development of western man's environment is now anarchic in just such a manner. Many of our great cities are rotting at the core, like ancient trees. Concrete arteries pump the life-blood of the organism to the commuter communities and out-of-town shopping centres on the perimeter. Twenty-, thirty-, hundred-storey tumours implant themselves, seemingly at random, in the system. Vast areas of land are blighted by the need to provide massed parking space. The sense of place is dissolving. Higgledy-piggledy over once beautiful towns spreads a creeper of wires and winking lights, poles, signs, hoardings, boxes and a multiplicity of man-made structures, smothering outlines, confusing the eye. Urban spaces are no longer pleasurable to move about in but imprison the spirit. Architecture becomes accommodation. Homes become living units. People become statistics, surrounding themselves with all the gimmicky affectations of the cut-price store.

Applied to any one spot, the picture may seem overdrawn. Here and there areas are developed or redeveloped imaginatively, good buildings go up, fine interiors are to be encountered. These, however, cannot disguise the underlying lack of coherence in the physical fabric of the world we are constructing as a whole, from Los Angeles to the new townships of Africa. It is a world held together by rapid communications and transportation; the trinkets on sale at its airports provide a suitable symbol of its design values.

* *The Classical Language of Architecture.* John Summerson. BBC 1963.

It is customary to ascribe the break-up of our design traditions to the industrial revolution. Certainly, for reasons which are familiar, the process was accelerated during the nineteenth century but its origins can be traced much further back. The general climate of taste in which an authoritative code of appearances becomes possible or otherwise has always been affected by certain interlocking factors. Some of these—the degree of specialization claimed by the artist and the degree of autonomy claimed for the work of art—group themselves round the status of the artist and his long fight for personal and aesthetic emancipation. Others—the use of materials and the power of tradition—group themselves round the time-scale on which society itself is developing and changing. A handful of random soundings will show how closely these factors have affected man's collaborative efforts.

Until relatively recent times, the artist was a workman, the commission was his starting point and excellence of craftsmanship his aim.

If the break between the artist as artisan and the artist as a channel of divine power is to be symbolized by any one moment of time, that moment may conveniently be held to be the completion of the Sistine Chapel in 1512. Thereafter the artist began, increasingly and successfully, to fight against anonymity for himself and to claim a degree of autonomy for his work; concurrently the idea of specialization spread through the arts and the crafts, finally to a degree hitherto unknown. It is rarely possible, as Sir Anthony Blunt has pointed out of building in France before 1525, accurately to define the function of any one craftsman. Exceptionally, workshop managers achieved social status, for example the mayoralty of York or Nottingham, but for the most part the artist remained a journeyman labourer. If we go back to ancient Egypt, he was little more than a superior slave, for whose calling the scribes seem to have had nothing but contempt. His training, it may be assumed from the plaster casts and other educational exemplars from palace and temple workshops, was closely controlled. The rigidity of the formalistic modes employed scarcely permitted of plastic invention, even had the demands made upon him been less static and unchanging in their nature. The skill of the artist-craftsman was concentrated wholly upon perfection of execution and the stereotyping of elements and patterns made possible a measure of specialization and rationalization of production which has been likened to that of the monasteries and the medieval cathedral workshops. The minor arts were controlled no less closely, so that a level of unified accomplishment was achieved so extraordinary as now to seem almost inhuman. The processes by which this magico-religious code, with its various renunciations,* was evolved are not now to be apprehended; it is nonetheless

* It must not be thought, for example, that the 'frontality' formula betokened an inability to depict the human body more realistically; there is evidence to the contrary in certain reliefs and plaster casts.

clear that, notwithstanding successive cycles of development in detail, for some three thousand years its practitioners were so completely conditioned and attuned to its general demands that they could conceive of no other. Such ant-like unanimity of thought will not be seen again.

In a cycle which was to be repeated two thousand years later, the classical world saw the artist's first advance to a more elevated status. Between the seventh and the sixth centuries B.C., Greece breathed life into the stone; in the culminating glories of the fifth, the work of art had dropped all but its outward claim to utilitarian or sacred purpose and had become an end in itself. The artist found for himself secular patrons—most commonly the city-state; most splendidly, after the battle of Salamis and the expulsion of the Persians, Athens herself. During the Homeric and classical periods his standing had improved but marginally: he remained an ill-paid journeyman who found it necessary to turn his hand to whatever offered. For long after that sculptor and architect—Polycleitas and Scopas may be instanced—were one and the same, like the material they used, but prosperity and the demand for their services greatly increased their renown. Pheidias was very famous—among his many tasks he was charged with general supervision of the amenities of Athens—and commanded supporting talent of the highest order. By the time of Alexander the Great, the artist's position had changed so markedly that it was possible for Apelles to become, not only court painter, but Alexander's confidant as well. And in the Hellenistic age that followed, emperors took to amateur painting, painters could make fortunes and artists of every kind could bask—erratically—in the 'discovery of artistic genius' which Schweizer attributed to the influence of Plotinus. The complete autonomy of the work of art, however; the final embrace of aestheticism; the ultimate release of the artist from any other responsibilities than to his own daemon—these were to be reached only towards the end of that similar cycle which began in the Middle Ages and culminated in our own day.

By the time of the monasteries there were aristocrats to be found among those painting and executing hand crafts; the monasteries themselves often undertook technical research in building matters and concerned themselves widely with craft work in metal, mosaic and design generally. Leo of Ostia recounts, in the *Chronicle of Monte Cassino*, how the famous abbot Desiderius set 'a great number of young monks' to learn from the Byzantine mosaicists brought to work in the abbey of St Benedict, 'so that their knowledge might not again be lost in Italy. And the most eager artists selected from his monks he trained not only in these arts but in all the arts which employ silver, bronze, iron, glass, ivory, wood, alabaster and stone.'

Not only supporting craftsmen but the master masons came from Byzantium and Italy. Even with the growth of the lodges, or companies, in the twelfth and thirteenth centuries, the artist was usually a stranger in the

sign is synthesis. A de-
n concept that has been
ally grasped and fully
lized is not to be broken
wn into its contributory
ments: their interpenetra-
n is complete. In the
cery and fan vaulting of
g's College Chapel,
mbridge, above, *structure*
d *form* are utterly fused.
are they, at a modest
el, in the Victorian cast
n balcony in Cheltenham,
ht.

Likewise, *form* and *decoration* are indivisible in the double arcaded hall of the mosque at Córdoba, begun late in the eighth century, above. Less exotic, but no less assured in its use of the basic structural materials for decorative purposes, the fifteenth-century flintwork of the Gatehouse of St Osyth's Priory, Essex, left.

In the total design concept, boundaries between the arts tend to become blurr Are the great Hindu temples of India—for example Perur, right, dating from the early eighteenth century—to be regarded as architecture or sculpture?

Such distinctions become academic in the face of overwhelming unity of aim, even whe it be expressed more subtly. The column above, from S. Vitale in Ravenna, is not only an exquisite piece of sculpture; it has affinities with the design of S. Vitale as a whole

The Palazzo Ducale of Urbino, constructe some 900 years later, is one of the marvels of its period. Federigo da Montefeltro, for whom it was built, was the 'complete man' of the Renaissance. The gilded ceiling of hi study, left, surmounts a series of remarkable *trompe l'œil* panels in intarsio, above which were portraits of Federico among his intellectual peers—the whole design, it is though having been inspired by Botticelli.

ɹst as great architecture is one with its interior detail-
g, so does it seek union with its setting. Two centuries
ɘparate the architectural formalities of the garden at
ugsburg, above, or the marvellous yew-lined
ooms' of the gardens of the Villa Marlia, near Lucca, below, from the 'organic' gesture of say, Frank Lloyd Wright's 'Falling Water'. The aim of total identification of building and setting remains the same. The contrived informality of the eighteenth-century English park, too, is merely a more subtle extension of the same relationship.

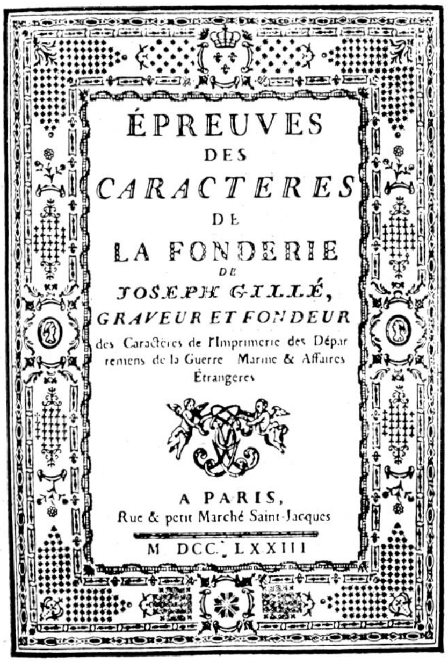
ÉPREUVES
DES
CARACTERES
DE
LA FONDERIE
DE
JOSEPH GILLÉ,
GRAVEUR ET FONDEUR
des Caractères de l'Imprimerie des Départemens de la Guerre Marine & Affaires Étrangeres

A PARIS,
Rue & petit Marché Saint-Jacques
M DCC. LXXIII

At times, when man's aspirations and abilities have coincided at hig level, all these things have come together in complete stylistic unity; every activity is expressed in a common language of form and ornament, so that such periods seem almost to result from a kind of collective compulsion.

Between 1715 and 1745 in France, a little later in Central Europe, furniture, decoration, porcelain, metalwork, even type design were imbued with the asymmetrical scrolls and counter-curves of Rococo. Above, engraving of the *boiseries* at Rambouillet; left, title page of Gillé's type book. Watteau carried the same spirit into painting. The churches of Bavaria, above all, carried it into architecture. The pilgrimage church at Wies, right, by Dominikus Zimmermann—an exact contemporary of Bach, Handel and Scarlatti—absurdly, brilliar and beautifully brings painting, sculpture, architecture and interior design into harness to give an exuberant foretaste of Paradise.

Within periods of stylistic harmony like these, it has been possible for artists and craftsmen to work as a team. About 1272 three members of the workshop of Nicola Pisano—we know their names: Lapo, Donato di Ricevuto and Goro di Ciuccio Ciuti—carved the marble above, as part of a screen for Siena Cathedral (it is now in the Victoria and Albert Museum).

No fewer than four painters—Antonio Balestra, Domenico and Giuseppe Valeriani and Giambattista Cimaroli—worked upon 'The Apotheosis of William III', right. Apart from such art factories as the Rubens studio, it was not uncommon for painters specializing in one subject—portraiture, animals, seascapes or landscapes—to be called in to collaborate on a complex work. In our own day, the early surrealists used to collaborate in the production of group drawings known as 'The exquisite corpse'.

Day-to-day and inch-by-inch collaboration was possible because architects and designers, artists and craftsmen worked within a common discipline and accepted a common aim. Developers, architects and builders were content, for example, in Georgian England to accept the anonymity of rationalized production methods and the standardized designs of the pattern books, preferring to concentrate their energies upon the larger unities. Six hundred joiners and cabinet-makers were among the subscribers to Sheraton's Drawing Book; right, a doorway by William Pain from *The Carpenter a Joiner's Repository* (1778).

For centuries, moreover, accord between successiv generations was well-nigh ensured by a language—th Orders; by a common scale of construction; and by t use of traditional materials. From Venice to Corsham (above), it was hardly possible for even the clumsiest builder to destroy a town's charmed continuity.

Frontispiece
3 f.t 9 in. into 4 parts, one is the diam.r of the Column, which is to be divided into 60 parts and these parts disposed to the moulding as figures in the following Plate wherein the measures agree with these on this Plate, but to a larger Scale.
1 diam.
W. Pain del.

The passage of time should not blind us, howeve to the *dis*unities of which the past was capable. How many of today's planning committees would accept the irregular fenestration of the Doges' Palace, below, without comment, or the access door tacked on so asymmetrically to the clock in Nuremberg (and with so little reference to the string courses), should such applications come before them?

By the same token, gratification at having brought building and sculpture into juxtaposition in the twentieth century should not blind us to th higher unities. Epstein's Madonna is in impeccab relationship with its setting; the absurd support t the Barbara Hepworth, a few hundred yards away destroys any sense of joint relationship between sculpture and building.

place where he worked, part, however, of a collective unit bound by precise codes of conduct and rights. In the monasteries the big workrooms, the *scriptoria*, were divided up among the specialists: the painters, calligraphers, initial painters and so on; and this trend towards co-operative specialization was given new impetus by the lodges. The team was led by the works manager, the *magister operis*, who looked after materials, labour, pay; and by the master mason, the *magister lapidum*, who was the co-ordinating architect and designer. The whole code of practice of the lodges reflected a will towards unity of spirit. Detailed rules were established governing the recruiting, training and payment of workers. Tools were provided, as were sleeping and eating quarters, and the lodges were able to offer a considerable measure of security while work continued in a particular locality. They also established the hierarchy within the team, restrictions on their members' claim to what we would now call copyright, and in general the subordination of the individual to the requirements of the job and its overall co-ordination.

Nonetheless one can sense here, for the second time in history, stirrings of aesthetic independence. It has been estimated that 25,000 names have been established in connection with the historical monuments of the Middle Ages —not all of them, it is true, designers and craftsmen, but more than enough to explode the romantic myth of anonymity. The changing status of the artist-craftsman is well seen in the differences between lodge and guild. In the fourteenth century, painters and sculptors began to withdraw from the lodges and set up in their own right. They then joined the guilds which, at least to begin with, were associations of equals and essentially defensive and protectionist in spirit. Their qualitative regulations—drawn up with at least as sharp an eye to the interests of their members as to those of their clients— unlike those of the lodges, allowed their members freedom of judgement in the aesthetic interpretation and execution of the brief.

Work on sculptural and decorative details, which hitherto had been executed *in situ*, began to be done elsewhere, away from the building site. Away from the site the specialist assumed a new authority. He began to find himself called in as a consultant by the communes and the guilds for their commissions and competitions, as, on a famous occasion, for the bronze doors of the baptistery in Florence. By the mid-fifteenth century, Alberti was counselling the architect to play hard-to-get.

> You should not immediately run and offer your service to every man that gives out he is going to build . . . I know not whether you ought not to wait till you are more than once importuned to be concerned. Certainly they ought to repose a free and voluntary confidence in you, that want to make use of your labours and advice. Why should I offer those inventions which have cost me so much study and pains, to gain perhaps no other recompense, but the confidence of a few persons of no taste or skill?*

* *On Architecture*, Book IX, Chapter XI.

The emancipation of the artist from his servitude to his employer has begun. He is beginning to hanker after self-expression. 'To run up anything that is immediately necessary for any particular purpose, and about which there is no doubt of what sort it should be . . . is not so much the business of an Architect, as of a common Workman.'* Filarete suggests that all artists should sign their work; the collector appears on the scene, and the first dealers. The Poggi case broke the grip of the guilds† and with Michelangelo the concept of genius as a God-given gift, which puts its possessor apart from the rest of mankind, was finally confirmed.

The artist, on the other hand, was still forced to apply his energies over a wide field. Though Vasari considered hand crafts somewhat *infra dig.*, we know from Nevi di Bicci that armorial bearings, shop signs, flags, designs for carpets and tapestry, tarsia work were executed in the painters' studios; wood-carving, metalwork, tiles and cartoons for this and that were among other things tackled in busy studios. Alberti lists among the things that the architect can give us: 'places for exercise and for swimming, baths, vehicles, mills, time-measures, trophies, tabernacles, sacred edifices, churches, rocks cut, mountains bored through, valleys filled up, lakes confined, marshes discharged into the sea, ships built, rivers turned, their mouths cleared, bridges laid over them, harbours formed, engines and machines of war, fortresses and the like inventions.'‡ Indeed, the very status of the artist served on occasion additionally to blur the underlying trend towards greater specialization. Giotto was elected master mason to the city of Florence entirely because of his eminence as a painter. Far from the centres of the Renaissance, if not always for the same reasons, things were not very different. John Blackshaw, who carved the statues of King and Founders for Wadham College in 1612–13, also built the barrel vault of the kitchen.

Specialization, clearly, did not come overnight. Nonetheless by the end of the Quattrocento the process of dividing and subdividing the job, and subdividing it again, had gained impetus and, once started, was not to be halted. Particularly in the craft world, its momentum has continued to increase steadily ever since. By the early seventeenth century the ordinances of the newly founded Clock-makers Company specified 'Clock-maker, Mathematical Instrument-maker, Sundial-maker, Graver and Case-maker' as branches of the trade. A century later, its recognized divisions included the movement-maker, the cutter; the pinions would be drawn at the mill; the springs would be made by one craftsman, the chains by another, caps and studs by a third; a silversmith would make the case; the dial-plates would come from elsewhere. All would be passed to the finisher, after the brass

* Book IX, Chapter X.

† A Genoese guild sought to enforce its prescribed seven-year course of instruction, but guild statutes were held not to be binding upon artists who did not keep an open shop.

‡ The Preface.

wheels had been gilded by the gilder, and whose name should decorate the face of the clock but the vendor who had employed these different tradesmen to put the pieces together.*

The personal authority and standing of the artist, and the diminishing area of activity on which he has focused his creative effort, have clearly changed the nature of any collaborative enterprise on which he may embark. So, too, has the time-scale upon which he works. When Sir Edward Pytts wished to build himself a house in the late sixteenth century, he bought the manor and castle of Kyre, in Worcestershire, in 1586. He made it his first task to acquire the brick and stone necessary for building. Two years later he obtained his first plan (at a cost of 40*s*.); master mason and workmen were on the pay roll (at 7*d*. a day). By 1592, chimney-pieces were under consideration; by 1595 the house was part complete. Sixteen years later, new plans were being sought, and on Pytts's death in 1617, Kyre was still not complete, though £2000 from his estate was earmarked for further work. The house grew, that is to say, organically over more than thirty years. The Kyre records, writes Sir John Summerson, 'underline the impossibility of attributing any house of the period to any one person. These houses grew under the free dictation of their builders, suggestions and decisions coming from various quarters at various times.'†

A building like St Peter's, or Gloucester Cathedral, constitutes a discussion across the centuries between architects and designers far distant from each other in outlook and taste. No one man, no single team, was called upon to envisage, from scratch, a complex whole in all its detail. The theme having been stated, elaborations, variations, decorations could be worked upon it by successive generations, not only in the light of changing requirements but in the full knowledge and understanding of the building itself—not a model, but the building as lit by the summer sun and the lowering skies of winter. How much easier to judge with precision and finesse the propriety and exact inter-relationships of those elaborations, variations and decorations. Even one man could take his time. Wren could spend twenty-five years building St Paul's before having to make up his mind about the shape of the dome.

Taste changes but, until recent times, it changed slowly, steadily, organically. Apprentice succeeded assistant, assistant succeeded master. Each generation inherited a framework of references, the greater part of which would still be in existence, still recognizable, at the end of its life. It no more occurred to the artist or designer that the creative act was thereby inhibited than it would occur to the football player that personal integrity demanded he change the rules of the game. And since the same syntax of appearances was

* From Campbell's *The London Tradesman*, 1747.

† *Architecture in Britain, 1530–1830*. Penguin Books.

accordingly common to all, it was possible for the designer to leave much of the detailing and decoration to the executant craftsman, in the same way that the eighteenth-century composer was content to delegate the appoggiaturas and other embellishments to the performer. Front elevations for Strasbourg and Cologne have survived but no comprehensive working drawings were made for the great medieval cathedrals. Templates and setting-out diagrams were kept by the guilds; beyond that a sketch and a set of measurements sufficed to guide the craftsman. Centuries later, Nash's designs for the streets and terraces between Regent's Park and St James's were perfunctory in their detailing, the working out being left to a host of builders and developers. When one of these, by accident or design, doubled the scale of the cornice mouldings for Gloucester Terrace, Nash observed only that they had 'come out larger than he expected'. Thus, we may imagine, did many of the architectural masterpieces of Europe come into existence.

Design, after all, no more than echoed the requirements of society and these did not change markedly in the short term. A domestic building of 1800 had to cater for much the same family needs as one of 1700, as did most public buildings. The concept of building for ephemeral needs, flexibility of internal planning, and a limited life, was unknown in the past. Thus could the London house between 1670 and 1870 adhere to a standard layout; thus could many of its components be mass produced; thus could they be assembled by less than skilled designers as a matter of routine with an appearance of mutual co-ordination. The pattern books of design—by Batty Langley, Sheraton, the Pains and the rest—which were such a feature of eighteenth-century building, were made possible only because the need they fulfilled was relatively static. A measure of elevational control was indeed exercised by Building Acts and, in the interests of uniformity, more demandingly by the estates offices of the great landowners (for example, on dimensions for different storey heights) when dealing with their numerous main and subcontractors. Essentially, however, the repetitive effects through which the Georgian street and the Georgian interior attained its character were acceptable to the public because novelty as such was not considered desirable.

The very materials used were unchanging. For thousands of years brick, stone and timber remained the friendly, familiar elements of building construction; with metal, pottery and glass they formed the basis for all three-dimensional design, from carriages to candelabra. Their properties, and the craft processes by which they were shaped, were as familiar as the laws of gravity. This unchanging material basis all but eliminated the imponderables of execution and performance and made it possible for the designer to concentrate wholly upon refinements of, or extensions to, previous solutions to comparable problems, rather than a totally new solution *ab initio*.

Small wonder that, until recent times, design canons and systems did duty for long periods with only small modifications. It is perhaps possible to distinguish between those based on some intellectual discipline—for example, the Orders—and those reflecting purely stylistic effects—such as Rococo or Art Nouveau. Here, in these latter, can be detected the quickening pulse of fashion. Once so weak that we used quite other, generic terms, like 'Byzantine tradition', by which to denote them, the imperious oscillations of taste today lead us to talk of 'schools' and 'movements' which may sometimes be of only two or three years' duration.

The physical and social turmoil of the nineteenth century was great, but no greater than the upheavals in its beliefs and its culture. Never before had so many of society's assumptions come under scrutiny simultaneously; never had so many threads of continuity snapped at the same time. Fanned and scattered by the gale of 1789, the sparks of the French Encyclopaedists found ready tinder and lit the way to the liberation of every mode of thought: leading in turn to romanticism's unequivocal premium on the uniqueness of personal vision, to the wilful obscurities of the aesthetic revolutionaries who followed, and finally, in our own day, to what Ozenfant has called an 'anarchy of egotism'. 'Schools came to birth, died and were reborn', said Valéry, on the occasion of his reception into the Académie Française in 1927, of the Symbolists of forty years earlier. 'Nor will I pretend that a desire to shock or break with tradition was entirely absent . . . nothing less than total liberty for the various forms of art and their expression was demanded.'

Excess, in fact, became praiseworthy. The heroic act of breaking ground untrodden by others was intensified into a compulsion for novelty at all costs, to achieve which the artist focused upon one narrow aspect at a time the creative energy that had once been applied to a wide panorama. Thus was achieved an intense impact of a more and more limited and specialized kind.

At the same time the virtual elimination of communications barriers, to the point where the transmission of ideas is all but instantaneous, has meant, in the first place, a speeding-up of the processes by which taste is formed and changed; secondly that cultural trends and spheres of influence are now world-wide. Initiation, acceptance, reaction, contempt and rediscovery now follow one another so rapidly that it will not always be easy for future generations, looking down the perspective of history, to disentangle the primary from the secondary and subsequent manifestations of an idea—for example neo-Dadaism from Dada itself. Where once the apprentice took his place, so to speak, in the queue, waiting to add his own small contribution to the sum of human knowledge, today's generation of students seems unbearably old-hat to its successor three or four years later. Art is traded in annually, like the Cadillac, and often designed, no less than the motor-car or the pop disc, for obsolescence. Tinguely's self-immolating doomsday machines come to mind, and fire bells ringing on their way to deal with burning 'skoob towers'. Compare, too, the marketing of cardboard throw-away furniture and such trinkets as the 'something box', America's Christmas gift idea of a welded steel box with eight tiny lights which winked at random for a year, after which they could never be made to work again. In short, as we all know, the restlessness and rootlessness of modern society calls for fashion kicks as a continuous process.

The changes wrought specifically by the turmoil of the industrial revolution were perhaps two. New materials and new methods of mechanical production led to a divorce between art and craft such as had never before existed; while the technological chain-reaction then set in train has led to the need to design structures, objects and images without precedent: telephones, typewriters, traffic signs, aeroplanes, electric razors, television sets, vending machines, animated film graphics and the rest. Indeed, it has led to totally new media of communication, and thereby to new art forms, which have tended to drain off some of the creative energy applied to certain traditional forms in the past. The television play, for example, has largely replaced the short story. Secondly, migration to the city, the rise to economic influence of an uncultured bourgeoisie and the gradual emancipation of the labouring classes, brought radical changes in the relationship between the artist and his client, and, at least in Victorian England, a rapid enfeeblement of all the arts—the visual and plastic arts in particular.

Add to these pressures those historical trends already noted and the collapse of a common language of appearances can scarcely occasion surprise. High art went one way, popular art another. In music the 'lascivious' waltz marked the watershed of what had been a single broad stream: in one direction lay jazz and the Beatles; in the other, serialism and *musique concrète*. In matters of design, the artist's preoccupation with his own aesthetic integrity—'Art for Art's Sake'—coupled with the diminished area on which the craftsman was focusing his creative effort, combined to pull both apart from that central position they had once occupied as one and the same person. Today, painting and sculpture have become so narrowly specialized, within so wide a total spectrum, that collaborative effort between artists is almost a contradiction in terms. John Evelyn wrote of Bernini that he 'gave a public opera wherein he painted the scenes, cut the statues, invented the engines, composed the music, writ the comedy and built the theatre'. Today, the dream of the universal man has gone, receded into a golden age as lost to us as any mapped by Claude or Watteau. The compositor can no longer design a page of type and the typographer cannot illustrate his text. If a painter turns to ceramics or takes his place in an architectural team, if a sculptor makes jewellery or a furniture designer a film, that is news and to be wondered at. We have destroyed utterly the kind of cumulative joint thinking made possible in the past by means of a canon held in common between artists, architects, engineers, craftsmen and designers. Henceforward, if the disciplines are to work in harmony, new patterns of collaborative effort will have to be found.

Before such collaboration can again become possible, two essentials must exist: a means of communication that is understood by all concerned and the will to use it selflessly. As has been seen, many intellectual and emotive concepts have, at different times, provided the basis for an acceptable canon: from mathematics to the greater glory of God. Some, it seems likely, were accepted by the artist all unconsciously. Some at least, like neo-Classicism with its coherent framework of ideas and stern ideals, arose by a deliberate effort of will and a consciously accepted philosophy of aesthetics. Conscious or unconscious of the controlling framework within which he was working, the artist was not, however, until very recent times, *self*-conscious about it. This very self-consciousness of modern man inhibits his readiness to subordinate himself to the relative anonymity of group effort.

Over the last eighty years, as will be seen, a general *rationale* of design has been established which is acceptable internationally and which does, potentially, provide an adequate means of communication. What is lacking is the will towards, the training for, and the habit of, inter-disciplinary collaboration. There is a horizontal unity between designers in a particular field, often all round the world, but rarely a vertical unity between different types of designers.

Only the artist himself can change this. There have been spectacular attempts, even in recent times, forcibly to impose upon artists and designers a common canon. The most successful was the great state machine set up by Le Brun and Colbert, who contrived not only to bring into it architects, painters, sculptors and craftsmen of all kinds, but to ensure that their design concepts were regulated according to the régime's doctrinaire principles.

As in Russia in our own day, seventeenth-century France attempted by committee discussion and majority vote (or sometimes by reference to referee) to formulate with clarity and precision, for their own and posterity's use, the ultimate aesthetic principles. The *Conférences* lasted from 1664 to 1674. The principles thus registered comprised an additive canon, as firmly and as uncompromisingly as the decisions and precedents which form the basis and body of our common law. Alas, though it has been said that the dividing line between the work of art and its frame was here abolished, this was also the beginning of academicism as we know it.

Political dictatorship will remain with us, doubtless, on one part of the globe or another, for some time yet. In our own more open society, however, one of the most marked characteristics of the second half of the twentieth century is the general lack of respect for cultural authority. Traditionally, the standards of the cultural élite spread gradually outwards, much as the gentleman's formal wear of yesterday becomes the waiter's formal wear of today. The young now show a fine disregard for their elders, however. In their clothes, in their styles of entertainment, in their values, they have cut clean and taken their own initiatives. Professionalism is mistrusted. There is a cult of the amateur. A tyranny of the aristocracy, of any cultural élite, has become a contradiction in terms. It seems impossible, that is to say, that conditions will ever again arise in the West whereby a drive towards aesthetic unity could arise from any other quarter than from the artists themselves. But before this can happen, humility must be relearned.

It is tempting to see in the Gothic cathedral a challenge of a kind the modern world cannot provide: in medieval man's submission before God a subordination of the individual will that is no longer possible. The creation, however, of a new capital city, a great liner, a university—these seem no lesser challenges in constructional terms, and to embody ideals to which secular modern man can subordinate himself no less wholeheartedly than did his Christian ancestors. The future environment of mankind, the fabric of our daily lives, the grace and propriety of our civilization, are bound up inescapably with the readiness of all the creative disciplines so to subordinate themselves to the relative anonymity of group working in the service of society. That is what this book is about.

2. The New Tradition

The nineteenth century has by now lost the black-and-white look of the immediate past. Through its coarse flamboyance and all its eclectic confusions the nuances can more readily be sensed. The table of its monuments is by no means yet definitive but few would deny the vigorous authority of its industrial buildings—its mills, its warehouses, its railway stations—or the bold originality of its engineering structures—its bridges, its glasshouses, often its very machinery. Though the nineteenth century marked the conclusion of a cycle of Western culture that had lasted for eight hundred years, inevitably it contained too the promise of what was to come. In painting, France held the key; in music, central Europe; in design, Britain.

Within the confused ideologies of the century, three distinct growing points of opinion may be detected: an odd trio, in some respects contradictory yet overlapping to a degree which made possible the eventual emergence of a new aesthetic. They were: moral revulsion at the social squalors of the era; a romantic longing to put the clock back to a golden age before the coming of industry; the arrival on the scene of the first new building materials since brick—iron, steel and, at the end of the century, reinforced concrete. From the first came the sense of moral responsibility which has gripped the architectural and design professions ever since; from the second came the desire to re-establish an ideal unity between the fine and applied arts; from the third came the concept of truth to materials and structural honesty, through which could be seen the first glimmerings of a machine-age aesthetic based on a functional programme.

The idea of utility, as a factor governing design, began to be expressed explicitly.* As far back as 1833, in his *Encyclopaedia of Cottage, Farm and Villa Architecture and Furniture*, J. C. Loudon (1783–1843)—in general a follower of the picturesque movement—writes of a terrace of houses he designed at Abersychan:

> The most useful is unquestionably the most beautiful. When the question is between a beauty belonging to the expression of art, and a beauty connected with fitness for enjoyment, it is clear to us that the decision ought to be in favour of the latter. In judging a building merely by its exterior, unless we are intimately acquainted with its use, we can only decide as to its symmetry, or other architectural expression. To form a just taste in

* In opposition to the Romantic taste—compare Keats's 'Oh sweet fancy! Let her loose; Everything is spoiled by use'.

> architecture, therefore, it is as necessary to study all the different purposes to which the different parts of the interiors of buildings are applied . . . as it is to study the original or conventional beauty of lines and forms.

Pugin it was, however, who most forcefully introduced the proposition that a building should express its purpose, and that to do otherwise was an 'unworthy deception', the result a 'sham'. It was he, wrote *The Times* in his obituary in 1852, 'who first showed us that our architecture offended not only against the law of beauty, but also against the laws of morality'. Architectural skill, Pugin held, consists in embodying and expressing the essential structure; and not in disguising it by borrowed features*—a maxim that would not be faulted at the Hochschule für Gestaltung at Ulm today.

A year or two later Sir Henry Cole spoke at the Council of the Royal Society of Arts in these words:

> Of high art in this country there is abundance, of mechanical industry and invention an unparalleled profusion. The thing still remaining to be done is to effect the combination of the two, to wed high art with mechanical skill. The union of the artist and the workman, the improvement of the general taste of our artificers, and of the workmen in general; this is a task worthy of the Royal Society of Arts and directly in its path of duty.

From Pugin the thread runs to Ruskin, from Sir Henry Cole to William Morris; thereafter to join in the Arts and Crafts movement and, later, English 'free' architecture. In Ruskin and Morris aesthetics, ethics and politics became inextricably mixed. Against their hatred of 'the English Minotaur of lust for wealth' and 'the labyrinth of black walls, and loathsome passages between them, which now fills the valley of the Thames', they set visions of a London 'full of gardens and terraced round with hawthorn walks'. They looked back, single-mindedly in the case of Ruskin, ambiguously at times in the case of Morris, to the selfless, communal creativity of the cathedral builders in the days before the machine. They saw in the revival of handicrafts the solution to their belief that nothing worthwhile could be produced without the work itself having given joy to its author. Ruskin intended his scientific museum for Oxford, into which he poured his energies between 1855 and 1859, as a testimony to communal creative effort. Not only did the workmen form something of a guild-colony in Oxford, but he attempted to press into service the more prominent of the Pre-Raphaelites (thus opening the door to their own diverting group attempt of 1857 to decorate the Debating Hall, now the Library, of the then new Union building). That even Ruskin found the result disappointing does not affect the significance of the ideal.

Holman Hunt wrote afterwards, of furniture he had devised during the years 1855–7, 'when I showed my small group of household joys to my

* *True Principles of Christian Architecture*, 1841.

PRB friends, the contagion spread'. Madox Brown, impressed by a chair of Egyptian inspiration, had a similar one made. 'After this the rage for designing furniture was taken up by others of our circle until the fashion grew to importance.'* So it was that when Morris wished to furnish the rooms in Red Lion Square that he shared with Burne-Jones after they came down from Oxford, finding nothing that he considered appropriate to a painter's surroundings available commercially, he set to with his friends and made some chairs and a table. Though he never again, so far as is known, designed furniture himself, he started, in 1861, a firm—known within the circle merely as 'the firm'—Morris, Marshall and Faulkner, Fine Art Workmen in Painting, Carving, Furniture, and the Metals. He himself practised at different times, in addition to his painting and mural decoration, weaving and textile design, stained glass, design for wallpapers, calligraphy and illumination, book production and type design. He sought, that is to say, to re-establish control of the whole design process from drawing board, through the means of production, to the end result; and to bring together into a harmonious working relationship as many different skills as possible.

It must not be thought that such ideas were confined to England. By the last quarter of the century they were becoming common to forward-looking designers everywhere. In Auguste Choisy, for example, one can find both nostalgia, for the renaissance man for whom 'all matters touching form were [his] domain', and at the same time the conviction that, once a correct appraisal of the problem had been made, form would follow function in the light of the technical means available.

Nonetheless, it was in England pre-eminently that these concepts were effectively plaited together—more by instinct, let it be admitted, than by systematic thought—and carried forward into the new century. 'Have nothing in your house', cried Morris over and over again as he stumped the country, 'which you do not know to be useful and believe to be beautiful.' In his wake a considerable number of talented young artists and architects turned to the 'useful 'arts; Guilds, Schools and Societies for the hand craftsman were started; the 'Arts and Crafts' movement took its place, ironically enough, alongside 'Art for Art's Sake' as the most significant trend of the end of the nineteenth century in Britain. Joined with it came first glimmerings of the meaning of the great engineers' contribution to design: Paxton, Telford, Brunel; Brindley, Arkwright and Stephenson; the railway bridges and the big termini, the Crystal Palace, the London Coal Exchange so recently and so sadly demolished. Ruskin, for one, could find nothing to admire in the Crystal Palace but as early as 1851 Matthew Digby Wyatt, writing in the *Journal of Design* of the Crystal Palace and iron architecture, found that 'It has become difficult to decide where civil engineering ends

* Quoted in 'Pre-Raphaelite Painted Furniture' by Michael Archer, *Country Life*, 1 April 1965.

and architecture begins'. Nearly sixty years later Lethaby, echoing Choisy's view of Gothic as a 'triumph of logic', could draw the general design moral: 'Gothic architecture was developed not by any aesthetic view of the proportions, but by getting the nerved vault, the ramping buttress and the stone-barred window to do the utmost possible'. 'The method of design, to a modern mind, can only be understood in the scientific, or engineer's sense, as a definite analysis of possibilities—not as a vague poetic dealing with poetic matters.'

This is the authentic voice of the twentieth century. To Adolf Loos in Vienna, savaging all ornament as immoral, engineers were 'our Hellenes'. In America, Frank Lloyd Wright was heralding '. . . the Machine Age, wherein locomotive engines, engines of industry, engines of light or engines of war or steamships take the place works of Art took in previous history'. Some of this was by way of understandable reaction to the charged, hothouse sophistications of Art Nouveau; some a reflection of the general iconoclasm that was everywhere apparent in the new century.

Art Nouveau in fact marks the real watershed between tradition and 'modern' design; wholly backward-looking in some respects, it was wholly forward-looking in others. Neurotic to a degree in most of its decorative detailing, relying as a matter of course for that decoration upon handicrafts, the movement nonetheless exhibited qualities of structural honesty, a frankness in its use of materials, a readiness to re-assess traditional concepts of form and space, which were entirely unfamiliar at that time. In Mackintosh's Glasgow School of Art, the will to bring the parts into integral relationships of a new kind is unmistakable. By the same token, the movement as a whole strove to bring its every manifestation into a single community of design ideas. Mucha's posters, Liberty's fabrics, the Paris Métro, the furniture and utensils of the British designers, jewellery from many sources—all are recognizably of the same family and represent a genuine international effort, albeit a highly self-conscious one, to re-establish unity between the visual and useful arts. Such was the power of the times that even so pure a painter as Bonnard found himself designing tables and a desk (for Thadée Natanson of the *Revue Blanche*), screens and stained glass. All over Western Europe, the same thing was happening.

Art Nouveau was, however, but a cosy curtain-raiser to the first decades of the new century, which were to span a period of ferment in the visual arts the like of which had not been known before. The will towards new forms was everywhere apparent. All previous concepts came under re-examination; old habits and thought-patterns were systematically destroyed; new foundations were sought for an art of unparalleled purity and intensity. The pictorial image and its content were broken up into their constituent elements—colour, form, anecdote, emotion and so on—each to be explored and exploited in isolation. In their concentration upon ever more narrow areas of sensibility, the seminal movements—Cubism and Futurism, Dada and Surrealism, expressionism and abstraction—thus reflected the general drive towards greater and greater specialization. At the same time, some of these movements—for example Futurism and Vorticism, Dada and De Stijl—sought equally to break down the barriers between the arts by means of an all-embracing aesthetic.

Nor was the process confined to the studio. Everyday life began to change. The rewards of the technological revolution—electric light, the motor-car, the camera, the typewriter, the telephone—were beginning to filter through to the middle class. These were man-made things for which no precedent existed and their manufacture served to clarify the distinction between design—which is the preplanning of form; craftsmanship—which is manual skill; and painting and sculpture—which are expressive communication.

Van de Velde, in Belgium, was one of the first specifically to call for a new relationship between the arts: 'What I recognize to be . . . the crux of all the artistic endeavours of our time is a yearning for a new harmony and a new aesthetic clarity.' In an address on Van de Velde's fiftieth birthday, in 1913, the art critic Karl Scheffer said: 'What you have always in mind . . . is a great new unity of the arts in which each individual art acts in accord with the others and in which all things, the highest and the lowest, the greatest and the smallest, are pervaded by the same form principle.'

We are here at the real starting point of the twentieth century's effort to create a new design tradition. For it was Van de Velde who was called to Weimar to set up, for the Grand Duchy, what he called a seminar—a meeting ground for artist, craftsman and industrialist—out of which grew in 1906 the Weimar School of Arts and Crafts, which was in turn to become the home of the post-war Bauhaus. Like Morris before him, he had designed all the furniture and furnishings, the light fittings, the utensils—even gowns for his wife—for the house he built in 1895 outside Brussels. Now he introduced the workshop method into the Weimar teaching system and linked it to that concept of mass-production which is the *raison d'être* of industrial design.

Professor Pevsner has pointed out* that both Pankok and Poelzig anticipated Van de Velde in substituting the workshop for the studio—and indeed that the idea came anyway from London. Priorities aside, however, architecture and design were beginning to move in directions no less radical than those of the other arts. The process was continuous, and leads through to the present day, but the creation of the basic tradition took place between 1907 and 1927, themselves years of special significance. Within this period a cluster of dates around the end of the First World War mark notable intermediate stages in what was a rapid crystallization of thought.

* *Architectural Review*, 793, March 1963.

Between 1896 and 1903 the apparently unlikeable Hermann Muthesius was appointed as a supplementary Trade Attaché to the German Embassy in London, with the special brief to report back on English architecture and design. On his return to Germany he became the most powerful and influential propagandist for a new approach to design—for quality, for the value of form, for standardization. With him the initiative passed from Britain to the Continent.

In 1907,* as the direct result of a trenchant public lecture by Muthesius, the Deutsche Werkbund was formed—an association of architects, designers, craftsmen and industrialists that would be a 'rallying-point for those able and willing to work for high quality'. Through its conferences and its less formal meetings, its exhibitions and pilot projects, the Werkbund was to be, for a quarter of a century, a powerful stimulus to clearer thinking and higher standards in all forms of industrial and industrialized design. Its success led to like organizations being set up in Austria, Switzerland, Sweden and—in the form of the Design and Industries Association, founded in 1915—in Britain.

This same year, 1907, saw also the appointment as design consultant to the vast AEG undertaking (the Allgemeine Elektrizitätgesellschaft) of a figure to be scarcely less influential than Muthesius. This was the architect Peter Behrens. With responsibility for the appearance of everything the company built, manufactured or printed, he was in some sense the forerunner of such figures as Frank Pick and Adriano Olivetti. As they did not, he, however, had direct creative responsibility for these things through the design office which he headed. Through this office were to pass, around 1910, Walter Gropius, Mies van der Rohe and Charles-Édouard Jeanneret/le Corbusier.

The still new idea of a machine aesthetic received powerful rhetorical support when Marinetti fired the Foundation Manifesto of the Futurist movement in 1909. Futurism's euphoric worship of technology, proclaimed in this and subsequent broadsides through to the 1914 *Manifesto of Geometrical and Mechanical Splendour, and the Sensibility of Numbers,* undoubtedly infected the plastic arts generally—in Milan, Paris and London at least—with a new dynamism. The movement did not survive the war as an organized force. As Wyndham Lewis wrote later of himself and Vorticism, 'a bigger *Blast* than mine had rather taken the wind out of my sails'.

It was in Holland, relatively untouched by hostilities, that the next gesture towards an ideal unity in the plastic arts was to be made. This was the De Stijl movement, the creation to all intents and purposes of Theo van Doesburg. 'I abhor all that is temperament, inspiration, sacred fire, and all the attributes of genius that conceal untidiness of mind', wrote van Doesburg.

* The year, incidentally, in which Picasso painted his 'Demoiselles d'Avignon', the pivotal work which divides modern painting irrevocably from the Renaissance tradition.

'The culture of particular form is approaching its end', wrote Piet Mondriaan, the movement's pre-eminent painter; 'the culture of determined relations has begun.' Finding its roots in Cubism, in Futurism, in the theosophist philosophy of M. H. J. Shoenmaekers, De Stijl was determinedly non-figurative, methodical in its procedures, minimal in its means and generally collectivist in its approach to artistic unity. In contrast to most such movements, that unity was more apparent in the works of its members than in their personal relationships. So powerful is the stylistic link between the paintings of Mondriaan, Rietveld's red-blue chair and the Schröder house, Huszar's logotype for the group's periodical and indeed any elevations, drawings or paintings by Oud, van der Leck and the rest, that one looks for a close-knit circle of intimates. In fact many of the group never met (for example, almost unbelievably, Rietveld and Mondriaan); within five years or so none of the original founders were any longer effective members; and thereafter van Doesburg gave houseroom to a number of more nihilistic and romantic figures and groups from the international scene. Nonetheless, De Stijl, in the first, domestic stage of its existence, remains the most potent twentieth-century example of the establishment of a common aesthetic language in use by architects, painters, designers and typographers. Indeed, self-abnegation reached an almost transcendental point. Painting, wrote Van Der Leck, 'will—by expanding from individuality to universality—demand from the building the entire colour- and form-conception belonging to painting as its rightful domain. If the architects look out for a painter who will create the demanded plastic image, then the modern painter also looks out for an architect who offers conditions suitable to attain an essential uniformity of plastic arts by joint efforts.'* More absolutely, Mondriaan wrote:

> The problem could perhaps better yet be solved by ceasing to make individual paintings altogether! If those who felt inclined were to have their apartments furnished and decorated in the style of the New Representation, the New Representation paintings could gradually be dropped out . . . There will come a time when we shall be able to dispense with all forms of art as known to us today; beauty will then have ripened into the tangibly realistic. Mankind will not lose by it.†

If De Stijl provided the most coherent and recognizable exemplar, a more profound and lasting contribution to new thinking in design and education was made by that most influential of schools, the Bauhaus. Gropius was called to Weimar, on Van de Velde's recommendation and as his successor, in 1915, there to evolve his plans for reorganizing the schools after the war. The Bauhaus opened in its new guise in 1919. In its rather more than ten years of life, until closed by the Nazis, it contrived to become the catalyst of all those tendencies we have noticed. Above all it attained, eventually and

* De Stijl, I.

† De Stijl, III.

notwithstanding inevitable squabbles, that community of spirit between architects, painters, designers and craftsmen towards which Ruskin, Morris, Voysey, Van de Velde and their like had all been groping.

The foundation document of the Bauhaus, the first Bauhaus Proclamation, exclaims:

> The complete building is the ultimate aim of the visual arts. Their highest function was once the decoration of buildings. They exist, today, in an isolation from which they can only be rescued by the conscious joint efforts of all craftsmen. Architects, painters and sculptors must once more recognize the nature of buildings as composite entities. Only then will their works be permeated with that architectonic feeling which has become lost in the art of the salons. [And again]: Let us form a new guild of craftsmen . . . and together create the *Bau* of the future which shall bring all into one single integrated creation: architecture, painting and sculpture, rising to Heaven from the hands of a million craftsmen, the crystalline symbol of a new faith!

To begin with, aspiration led reality. The school was indeed a fusion of the Weimar Academy of Fine Art and Van de Velde's design and crafts school but the staff at the outset were mostly painters—Expressionists from *Die Brücke*—and the emphasis was on workshop procedures and handcrafts as a teaching discipline. Everyone was an 'apprentice' or a 'journeyman'. Together with the *Vorkurs*, or preliminary course that was intended to purify the incoming student of his accumulated preconceptions, this emphasis on the crafts formed the 'Bauhaus method' that was to spread across the world and become general for advanced design training everywhere. Briefly, the Bauhaus sought to develop the responses of the designer over as wide a field as possible, through constant confrontation by visual and tactile stimuli, while at the same time developing his ability to analyse a design problem in depth, to build logically and rationally upon the basis of that analysis. The latter part of the proposition only became really possible after the school's internal battles of 1922 to 1924: inevitable clashes between the intuitive Expressionists on the one hand and the socially minded exponents of rationalization, industrialization and standardization on the other. What was at stake was the establishment of a universally valid methodology around which could grow a new tradition. Feininger wrote in the autumn of 1922: 'If we can't win over the industrialists, our expectation of survival is low. Our aim must be to earn, to organize production, to work for quantity production. . . .' The following year van Doesburg's visits introduced the disciplined classicism of De Stijl and Moholy-Nagy joined the staff, rapidly to become the school's most influential spokesman. Inevitably, the Expressionists lost the battle—for expressionism, by definition, cannot be taught. One or two left, or transferred to other jobs, and henceforward the school exhibited a firm bias towards industrialization and rationalized design

e impact of the Industrial Revolution upon long-
:ablished patterns of society broke a craftsmanly
dition of centuries. At the same time, its
hnological consequences—in the form of new
terials with new properties—laid the foundation
a new tradition. First of these materials was
n, its inaugural monument Abraham Darby's
st iron bridge at Coalbrookdale in 1777, right.
ere followed Telford, Brunel and Paxton, the
cks and the soaring railway stations—below right:
rk—and the great glass houses and buildings
Oriel Chambers in Liverpool, below, in which
fabricated iron provides not only the structure
t the cladding panels as well. And after iron
ne steel, reinforced concrete and a whole
noury of plastics. The list will continue to
gthen.

tions to pages 46–7 appear on page 48.

WREN'S
CITY Ye
CHURCHES
BY
A·H·MACKMURDO, A·R·I·B·A,
1883
G.ALLEN, SUNNYSIDE, ORPINGTON, KENT.

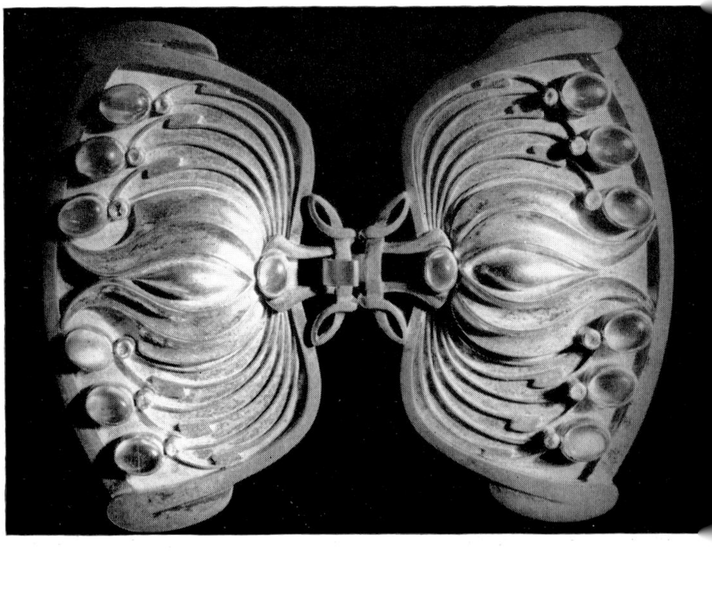

JOB
Mucha

Of the first attempts to re-establish an all-embracing style-language that would be independent of the past, two have a particular bearing on the modern movement and the creation of a new tradition: Art Nouveau (previous two pages) and De Stijl. Art Nouveau, like Rococo, was essentially decorative. However, it embraced within a common idiom architecture (the Métro, Mackintosh's Glasgow Art School), furniture and furnishings, jewellery and ornament, graphic design and to a degree—for example in Munch—painting. It flickered intensely but, lacking a functional or ideological programme, could not live. Page 46: scheme for an interior by Alphonse Mucha; title page by A. H. Mackmurdo; buckle by van de Velde; page 47: theatrical poster by Mucha.

e Stijl was essentially a
tional movement but, for
e years or so, it brought
the visual arts within a
mmon design hegemony.
much at one were the
tists of De Stijl that they
ked forward to a time
en painting would be so
ly integrated with archi-
ture that it would cease
have any independent
istence. The right angles
d primary colours of De
jl lived on in the Modern
vement as a whole but
fusion of aspirations re-
ined an aspiration only.
cing page: the Rietveld
-blue chair, second ver-
n (red back, blue seat,
ck frame, yellow ends to
frame members). Above:
mposition in red, yellow
d blue, 1921, by Piet
ndriaan.

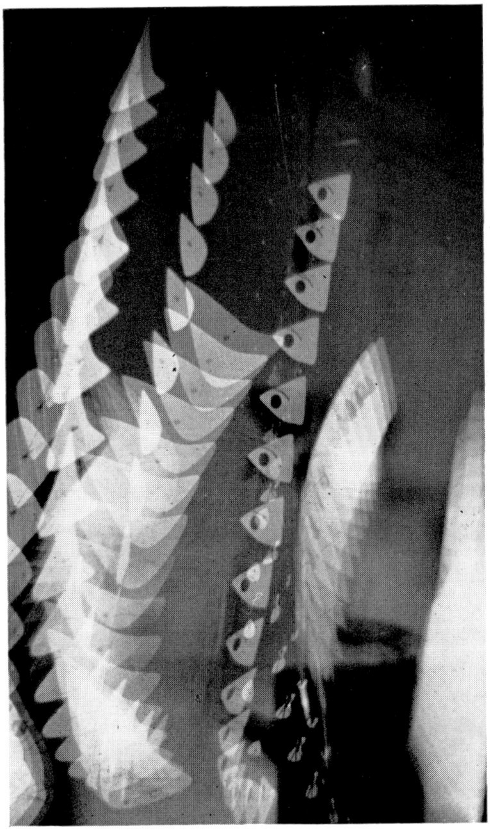

New materials and new styles were matched, increasingly, by new problems, for which no traditional solution existed. The three objects above had no parallel a few decades ago. Left, hairdryer by Ponti-Fornaroli-Rosselli; centre, traffic lights moc up by David Mellor on trial for the Ministry of Transport; right, Millward electric shaver and case by Kenneth Grange, which won the Duke of Edinburgh's Award for Elegant Desi in 1963. They could be matched by scores of others: televisi sets, tape recorders, typewriters, vacuum cleaners, Thermos flasks, flip-top packs and illuminated signs. To shape such things for easy use and quantity production, there came into existence a new profession: industrial design.

So varied are the problems that the modern designer has to solve, that no simple canon, however detailed, could be expected to embrace them all. The new tradition rests rather upon certain common assumptions and principles in the **approach** to the problem. Four in particular may be instanc

1. No form is so traditional that its radical reconsider tion is forbidden. Painting can include three-dimensional elements; sculpture can move (left, the tossing cascade of a Calder mobile caught by the stroboscopic camera). The traditional housing layout can be turned inside out on Radburn principles, so that the 'front' door is at the rear (to avoid traffic) and service access to the 'back' door is in fro (Farsta, near Stockholm, foot of facing page). Modern constructional techniques can free the rooms inside from the traditional distinction between wall, ceiling and roof: consid the Sydney opera house, Saarinen's TWA Terminal, overleaf, and the geodesic domes and space frames of Buckminster Fuller and Konrad Wachsmann.

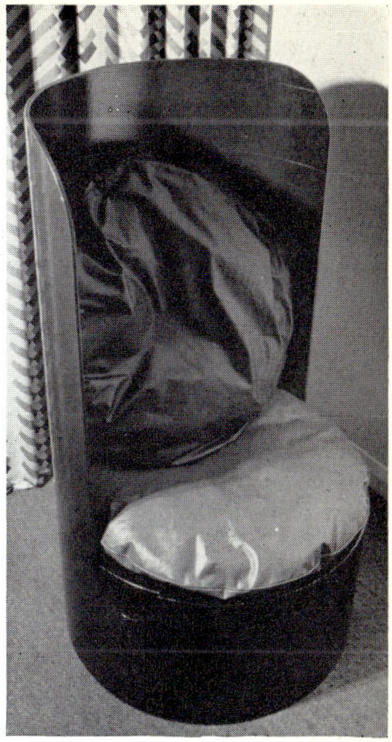

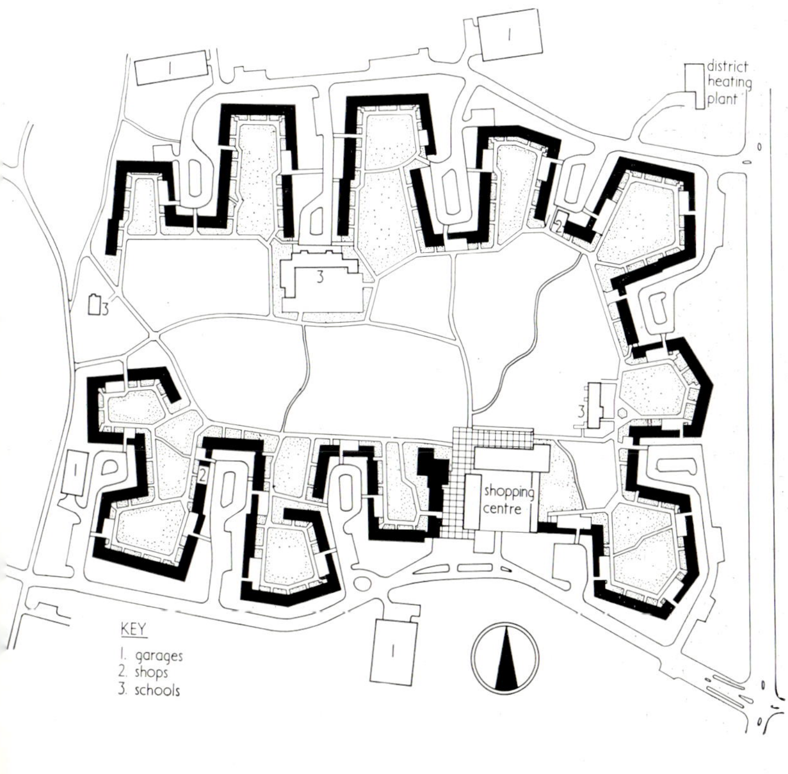

Alex Moulton had the temerity to query the divine right of the bicycle's 26-in. wheel, top of page; knives and forks have changed their shape for the first time for generations (this set, centre, designed by Alberto Rosselli of PFR). And need things be built to last? Why not paper clothes and throw-away furniture? The Tomoton expendable chair, right, designed by Bernard Holdaway, and made from compressed paper tube.

Terminal for Trans World Airlines at John F. Kennedy Airport, New York, by Eero Saarinen. See page 50.

2. Function, in any true sense, must embrace needs that are more than utilitarian. Drama, decor tion, acquisitiveness are but three of the many satisfactions for which man demands provision in his environment. The expression of his spiritual needs is part of the 'function' of a church: right, Notre-Dame du Haut, Ronchamp.

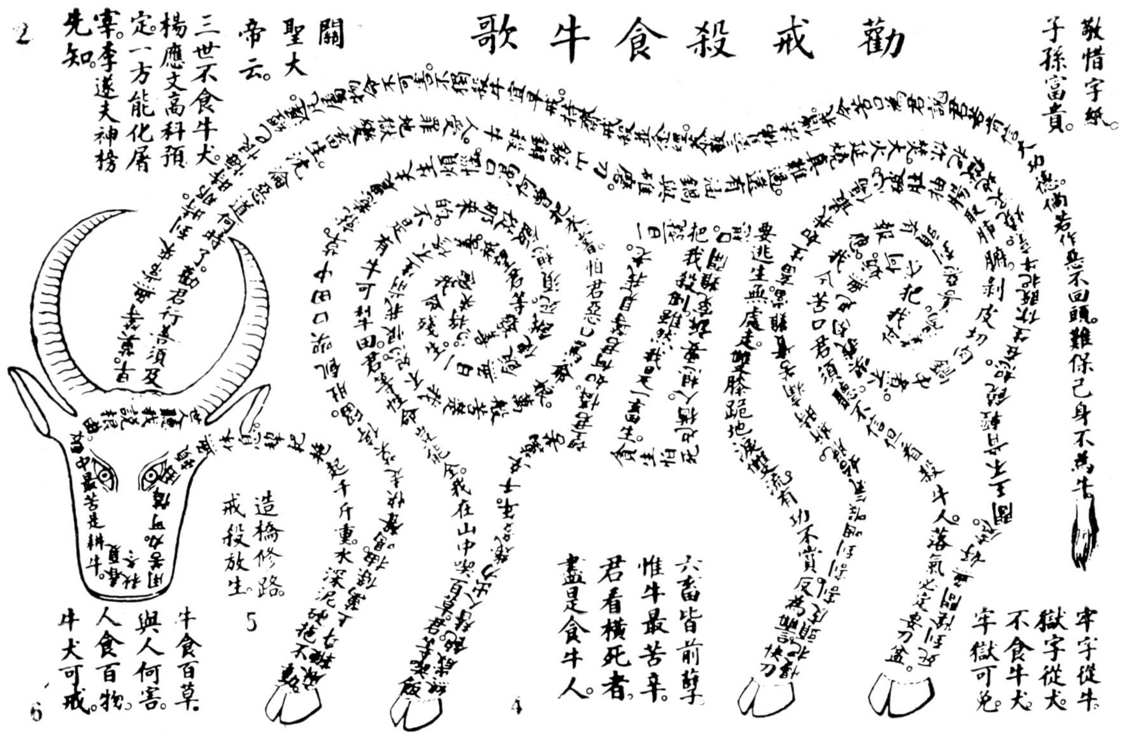

Nooks, crannies, shelves and spaces where he may display the particularities he has gathered about him as evidence of his god-given uniqueness: these are part of the 'function' of a house. Above right, a corner of Alexander Girard's house in New Mexico. *Pace* the transcendental purists of De Stijl, man craves colours other than the primaries; organic and decorative textures no less than inert surfaces: below right, wall relief by George Mitchell. To the pleasure of patterns he returns compulsively; in his typography he seeks wit, finesse and grace no less than legibility—above, Chinese exhortation against eating meat.

EXHIBITION

Closed solutions are invalid; open-ended, infinite ·lutions demand a sense of continuous space. ɜarly every aspect of design in the second half of the 'entieth century is marked by this sense of continuous ·ace. It may be seen in such a painter as Rothko; ubiqui-usly in sculpture as a result of pioneer achievements Pevsner, Moore and others of their generation; ove, infinity glimpsed through Moore's UNESCO clining figure. In architecture it has led to the wide-read use of *pilotis*, to the linking decks of Park Hill, ·effield, above left, and the vertical separation of ıctions at, say, Cumbernauld, or Philadelphia, or in ɜ Hook study. Exterior space and interior space ɛerpenetrate through patios and walls of glass; below t, Saarinen's headquarters for Deere and Company, ·ose products are shown on pages 125–32.

Inside buildings the same sense of interpenetrati
space informs every kind of interior, not only in
the disposition and linking of areas for different
purposes but in the disposition of the contents—
above, exhibition in the Cullinan Hall, Museum of
Fine Arts, Houston—and in the very detailing; lef
door in the Royal College of Physicians, by Deny
Lasdun and Partners.

4. The elegant solution is reached with the minimal means; structure, form and decorati are integral. Designers are struggling to feel the way, in modern materials for modern purposes, towards that same fusion of elements which marked the great periods of the past. Concrete is no longer rendered, as in the early days of the Modern Movement, but left with the textural impr of its form-work for decoration, facing page, top left. The brickwork of an acoustic wall by Hans Coper, above right, likewise forms its own decoration.

wo famous modern chairs, below, by Charles Eames nd Hans Wegener—one metal, one wood, but each a esign solution in which structure, form and decoration re not to be separated. Braun's little heater, right, has een much copied; its elegance remains unsurpassed—erhaps because none of its competitors have been so tally without pretensions.

In the hands of Nervi, the pre-eminent architect-engineer of the twentieth century, stress calculations and structural rationale, form and detailing, above, are as integral as they were in the Gothic cathedrals. The small stadium stand at Galashiels, by Peter Womersley and Ove Arup and Partners, below, is no less lucid ar authoritative. In these two structures, as in the Eames and Wegener chairs, the new tradition may be seen to be operating: not stylistically, but in a common approach to the problem.

programmes, based to a large extent on Moholy-Nagy's elementarism.* It was not, however, until 1927 that it actually set up an architectural department, the head of which, Hannes Meyer, took over from Gropius for two years when the school moved to Dessau in 1928. The Meyer spirit may be gauged from an open letter, 'My Dismissal from the Bauhaus, 1930:†

> What did I find on my appointment? . . . A 'university of design' which made the shape of every tea-glass a problem in constructivist aesthetics. . . . Everywhere art had a stranglehold on life. It became my aim to place design on a scientific basis. . . . The constructional engineer joined forces with the industrial consultant.

It is from these last years, notably in the person of Max Bill, that the direct link runs to the Hochschule für Gestaltung at Ulm which, in some respects, may be regarded as a mid-century equivalent of the Bauhaus.

Two other events in 1927‡ were of no less significance. The show houses of the Werkbund's Weissenhof exhibition, near Stuttgart, displayed the leaders of the modern movement shoulder to shoulder—Mies van der Rohe, who planned the exhibition as well as contributing a famous block; Gropius, le Corbusier, Oud, Behrens and Stam among others. It was here that Alfred H. Barr coined the tag 'international style' by which the movement was to be known for the next twenty years. Around 1926–7, too, three remarkable chairs were designed§—by Marcel Breuer, Mies van der Rohe and Mart Stam—all of which, by their use of steel tubing, jointly marked something of a watershed in furniture and industrial design.

The first of these two events symbolized the close-knit, almost family, relationship which existed between the leaders of the modern movement and the restricted network by which its ideas were spread. Gropius, Mies van der Rohe and le Corbusier, as has been pointed out, had all passed through Behrens's office at AEG around 1910, to absorb from him his belief in a hegemony of the visual arts, under the architect as universal designer. From him too they accepted, as a matter of faith, the virtues of and necessity for standardization. Converging on, and mingling with, this stream of thought as enshrined in the Bauhaus, was the controlled machine aesthetic of the De Stijl movement. From the Bauhaus the gospel was spread throughout

* 'Elementary is the art that does not philosophize, but is built out of its own proper elements alone.' *Aufruf zur elementaren Kunst,* reprinted in *De Stijl,* IV, and signed by Moholy-Nagy among others.

† Published in *Hannes Meyer* by Claude Schnaidt, 1965. Reprinted *Architectural Design,* December 1965.

‡ 1927, incidentally, was the year in which *Vers une Architecture* was translated into English.

§ The actual dates and priorities of authorship are not clear and perhaps do not matter very much.

Europe and North America by Moholy-Nagy; Gropius brought it to England, whence it spread to the Commonwealth, and to the United States; le Corbusier took it to South America, his pupils to Japan.

Of course the energy of the 'heroic period' spilled over into the nineteen-thirties. Le Corbusier's Pavillon Suisse of 1930, no less than his Villa Savoie at Poissy completed the following year, were seminal concepts that affected the thinking of a whole architectural generation. Both floated on *pilotis*, so that their main bulk was no longer rooted to the ground but reached up, like the shining towers of le Corbusier's *Ville Radieuse* set in their sea of trees, towards the sun and the sky. Implicit in both was the vision of a way of life in total contrast to the slums of the nineteenth century; in the Pavillon Suisse the vision of a city. The latter has been described as *the* academic building of the 'thirties. It could be held to mark the culmination and conclusion of the real heroic age of modern architecture.

There remained refinement and consolidation. In this a notable tool was CIAM—an informal group, largely fashioned by le Corbusier in 1928, of the leading architects and planners in Europe. CIAM remained in existence for more than thirty years but by the late 'forties the field work was done, the missionaries and the proselytizing were no longer necessary; the modern movement had become simply the way buildings were designed the world over.

The second of those events in 1927, the almost simultaneous appearance of the three tubular steel chairs, can be seen in retrospect as symbolizing the birth of a new profession: industrial design. The first step had been taken earlier, by Gerrit Rietveld, with his 'red-blue' armchair. The first version of this sculpture for sitting in was composed of four supporting and enclosing planes, two arm planks and a space-frame structure of square-section machine-cut rails, which never penetrated one another, as in the traditional mortise and tenon joint, but were dowelled so to speak in passing. The radical re-appraisal of the idea of 'chair' which this represented was the first application to furniture of that unencumbered, unprejudiced spirit of analysis that was central to the modern movement. It remained something of a research thesis, however, and it was rather the cantilevered tubular steel chair produced in such numbers during the 'thirties which marked the real breakthrough in terms of suitability for quantity production.

What, above all, the Breuer-Mies-Stam chair demonstrated was that there was a whole new field of smaller scale objects and structures awaiting the same total reconsideration that the architects were giving to building. Morris's vision of beauty and quality for the masses seemed a dream no longer. The whole of everyday life, all the things with which man clothes his body, his mind and his home, were seen to be worthy of the designer's attention; to perpetuate by mass-production anything less than well designed became indefensible. Of what use the new materials if 'traditional formulae hinder their proper use'? 'Instead of trying to solve new problems with old forms, we should develop new forms from the very nature of the problem.'*

Into this vast field came, stumblingly, a new kind of designer. Whereas most of the names so far mentioned in this chapter have been those of architects, now other talents applied themselves to designing for industry. To take four names which emerged in the United States during the 'thirties: Norman Bel Geddes and Henry Dreyfuss came from the theatre; Raymond Loewy was an engineer; Walter Dorwin Teague a typographer. Working for the most part empirically and *ad hoc*, men like these began to mop up the areas of design not covered by architects: appliances and consumer goods, furniture and interiors, transport and heavy equipment, exhibitions and visual communications.

Since, by definition, they were concerned in one way or another with quantity production and mass selling, their work was experienced directly by a wider public than ever, between the wars, saw a 'modern' building or an exhibition of 'modern' painting. Designers thus became one of the main channels by which the whole complex of ideas associated with modern architecture, modern painting and modern sculpture came to be assimilated by the general public. It has often been pointed out that the multitude is prepared unquestioningly to accept in advertising a visual idiom by which it would claim itself baffled on canvas. A new poster by A. M. Cassandre or E. McKnight Kauffer was an event in the nineteen-thirties, a talking point, and not only for those who 'did' the Bond Street galleries and the London Group. So, too, with objects about the house. For Mary McCarthy's Vassar girls of the early 'thirties,† everyone thought Russel Wright 'quite the thing now; he was using industrial materials, like the wonderful new spun aluminium, to make all sorts of useful objects like cheese trays and water carafes'. As recognition came to the designer, he became aware of the novelty of his position and sought to formalize it in professional terms; the Society of Industrial Artists and Designers, the senior body of its kind, was formed in Britain in 1930.‡ As his professional status grew, so did recognition become more formalized; in 1936–7 the Royal Society of Arts created the faculty of 'Royal Designers for Industry', limited at any one time to seventy members. Periodicals on design appeared in many languages. World Fairs employed the designer to fashion and clothe them and gave fresh impetus to his activities. The war made many and varied demands upon his services, and with the peace he found himself the object of governmental concern. National agencies—the Council of Industrial Design in Britain was

* Mies van der Rohe, in Bruno Taut's magazine *Frühlicht*.

† *The Group*, published by Weidenfeld and Nicolson.

‡ Revealingly, as the Society of Industrial Artists; compare *esthéticien industriel* in France and *Formgeber* in Germany. Only since the war has the profession sufficiently found its identity for the term 'industrial designer' to have received general recognition.

the first of its kind—were created with the express purpose of stimulating higher standards of design. National and international awards—for example, the Rinascente's *Compasso d'Oro* and The Duke of Edinburgh's Award for Elegant Design—were initiated. Schools and their curricula were completely reorganized. The professional organizations formed their own international federations—ICSID (The International Council of Societies for Industrial Design) and ICOGRADA (the International Council of Graphic Design Associations). Designer-manufacturers, or designer-manufacturer teams, have greatly increased the flow of well-designed furniture to the retail market and, for the European public at least, the greater freedom of international trade has meant wide distribution of distinguished furnishings and household equipment from other countries. A chair by Eames or Bertoia or Jacobsen has become as much a prestige object in the home as was a Chippendale procured at a country-house sale to a previous generation.

Design has in fact been but one aspect of that great explosion of popular enthusiasm for the visual and plastic arts which has proved such a phenomenon of the post-war years. All the way from soaring attendances at art galleries and the runaway sales of art books, to a steadily increasing concern with environment and physical planning, there are signs that society is once more beginning to use its eyes. As yet, it is not very practised, however. Popular confusion between 'styling' and design is as profound as popular acceptance of any new style of painting is uncritical. More good—and even distinguished—buildings were put up in Britain between 1955 and 1965 than during the whole of the previous half-century; but the damage done to our urban environment during the same decade was probably greater than in any previous decade.

For designers themselves, the very acceptance of the concepts which the pioneers proclaimed has brought with it a diffusion of their predecessors' aspirations.

> The notion of a new architecture [wrote Colin St J. Wilson recently],* was always a polemical one in which, for better or for worse, a whole body of ideas was at stake—social, technical and formal. In this body of ideas, all elements from door handle to city plan were so bound together that the form of a chair even project implications for the form of the city. ... Stakes of this order demand a kind of Hippocratic oath.

Today we have come so far from the ideals of the early part of the century that Dr Tomas Maldonado, Rector of the Hochschule für Gestaltung at Ulm, can declare that the designer must specialize; he cannot be 'universal' any more—except, he goes on to say, for a universal and vital curiosity in the whole human field.†

* 'Open Letter to an American Student'. *Program* No. 3, School of Architecture, Columbia University NY. Reprinted in *Architectural Design*, March 1965.

† 1965 Lethaby Lecture, Royal College of Art, delivered at the RIBA, 4 March 1965.

Philip Johnson says: 'Beauty is the only thing of any importance'.* Man is adaptable. 'If a chair is beautiful, you'll find a way of sitting in it.' And of painting and architecture: 'They have nothing to do with each other.' Another American, the sculptor David Smith, said of architects and sculptors: 'There are no affinities between us.'

By the middle of the century, the whole climate of aesthetic thought was moving away from definitive solutions. In a world of random particles and probability laws, of evanescence and rapid obsolescence, there is a mistrust of absolutes and ultimates. The dribbles and splashes of action painting, the return to the ready made and assemblage, the gay and melancholy flux of kinetic art, the throwaway encounters of pop art and the flickering ambiguities of 'op' art; the renewed weight given to theatrical and musical improvisation; novels of which the pages may be read in any order—these are symptomatic of the nineteen-fifties' and sixties' view of art as an attitude and a *process* rather than as the promulgation of perfected statements. The world of the architect and designer, a world of material problems and solutions, has to be less carefree. Inevitably, however, some of the dust of other forms of artistic creation brushes off upon the designer and Hippocratic oaths are currently not much in favour among the young. Characteristically, it was mistrust by the younger generation of what they regarded as the over-simplified codifications of CIAM that led to its decease and the formation of Team X—a group 'concerned to find the archetype in transformation rather than impose a prototype'.†

Yet it would be as mistaken to imply the total rejection of the tenets of the modern movement as to imply their universal acceptance. The formal qualities of the movement's heroic period were more often than not based upon 'a geometry of crushing banality' and frequently showed an egocentric disdain for the realities of life (who would wish to dust a houseful of Rietveld 1917 armchairs? or have to maintain the cracking, staining and weathering surface of a white concrete cube in a northern industrial city?). The vitality of its thinking, however, is not in doubt. The beliefs of its pioneers are with us still, if in amended form. The marriage they sought between art and industry has been largely achieved—not in the sense that all society's products are now adequately designed but in the sense that it is no longer exceptional for designers to receive industrial training or for industry to employ them. The rationalization and standardization of components sought by the pioneers is at last gaining ground. The 4-inch module for the building

* 'Monitor', BBC, December 1964.

† Time and time again new building types and forms demanded by newly emerging social functions—for example the airport, the tele-communications tower, the olympic stadium and multi-purpose sports hall, the motel, the multi-level car park—have been proved out-of-date almost before their completion. Hence such conceptions as Mr Cedric Price's Fun Palace—a structure designed for the utmost flexibility of use.

industry, the use of international 'A' paper sizes in printing, the development of industrialized building systems—splendidly demonstrated in one form in Britain's CLASP school at the 1960 Triennale and reinforced by the creation of the National Building Agency—the production of what may be termed modular ranges of typefaces, of which the Univers family is the most recent and most fully worked out: these are straws in the wind and indicative of what must come.

Design training of a richness that would have seemed unbelievable between the wars is now widely available in most Western countries and signs abound that the vision of an ideal unity between the arts plagues men's minds still. The best-known designers—for example, Ponti, Jacobsen, Wirkkala, Noguchi, Burle Marx, Charles Eames—move from one field of practice to another with no less ease and authority, *pace* Dr Maldonado, than their Baroque predecessors. The common bond linking all forms of design activity finds professional confirmation in the constitution of the Society of Industrial Artists and Designers, members of which practise in every field, from textiles to heavy engineering, typography to ceramics, television to exhibitions. Great undertakings in commerce and industry submit themselves to 'corporate identity programmes' with the object, as with AEG half a century before, of bringing all aspects of their activity into a common design framework. Great exhibitions have been created to demonstrate the possibilities of total design—from the Stockholm Exhibition of 1930, through the South Bank Exhibition of the 1951 Festival of Britain, to the Triennali in Milan of which three or four at least proved remarkably persuasive. Conspicuous efforts have been made, with varying degrees of success, to harness the allied arts in the design and decoration of important public buildings such as the UNESCO headquarters in Paris and Coventry Cathedral in Britain, while the central areas of new towns and some of the redevelopment of older ones represent honest attempts to create consciously, through teams of designers, the environmental values that in the past came about as a result of organic growth over long periods. 'Townscape', indeed, has become a subject for study in its own right and represents the growing awareness of design inter-relationships at its most significant level.

If so much has been achieved in a few decades, why does so much of our modern world remain ugly beyond belief? The reasons are often to be found in economic, political or administrative limitations. We are unwilling to spend enough, preferring to gamble; design management is so weak as to be well-nigh non-existent, so that boards and councils are inadequately advised; real environmental planning on a national scale is considered politically rash; and so on. Our culture as a whole is confused. In part this arises from the conflict between High and Low Art (though there are numerous indications that the gap between them is closing) but perhaps also in this: that creative energy has deserted certain traditional fields (for example the short story, portrait painting, book illustration) and has occupied others without precedent (the television script, the assemblage that is neither painting nor sculpture, the animated film title). This is no new process. Illumination was superseded by wood engraving, the fortified castle by the country house. Today, however, it happens very much more quickly and the natural gravitation of creative energy to the mass media—as the obvious way out of the artist's alienation from society—has undoubtedly led to a slackening of aesthetic tensions. Public response to the arts is widespread but too often based upon a glib confection of half-truths and *idées reçues*.

In matters of design, a not dissimilar diffusion of creative energy has been created by the Western world's explosion of industrial productive capacity and building construction. A poorly designed object may be produced by the million; an uncouth building may by its sheer bulk make an impact on its surroundings a hundredfold greater than anything in days gone by. The pioneers of the modern movement, as has been said, were a small band; much of their thinking had to remain theoretical and it was not difficult to keep their common aesthetic looking pristine and ship-shape. By the same token their influence was for long confined to a limited circle and their impact on practical affairs negligible. What has been sketched in here, that is to say, is not an objective picture of the overall, collective aesthetic of a century—it would not be hard to string together as many events and quotations having totally different implications.* The complete picture would present a far more complex, muddled and depressing image, the constant background of which would be made up of the persistent confusions inherited from the industrial revolution: confusions only slightly less numerous today.

What we have been following is the growth of a group of ideas, at first promulgated by a few men of vision in the first quarter of the present century, then widely disseminated throughout the world during the second quarter of the century, until now, weakened, distorted, modified, vulgarized, consolidated, extended, they affect in some measure every aspect of our daily life. The significance of those ideas is that they represented, and still represent, within the wider confusions of our society, the first reasonably coherent, consistent design language to have been evolved since the passing of the classical Orders and, indeed, one capable of application to an infinitely wider range of problems than were the Orders.

Some of the tenets governing the syntax of the language have passed into the vernacular as glibly as the terms of Freudian psychoanalysis: form follows function; truth to materials; structure must be expressed; and so on. Though still broadly acceptable, these no longer really serve adequately to explain current architecture and design. Nonetheless, they at least make clear one vital fact. The pioneers were emphatically not setting out to forge a *style* but

* For example, French provincial furniture in the style of Louis XV was still being made by country craftsmen as part of a living tradition, right up until the First World War.

a process and a set of principles. In practice, it is true, they deluded themselves that some of their stylistic trappings—flat roofs and white rendered walls—were the outward expression of functional and structural principles, but subsequent divergences of a national and personal kind have served to rub in that the connection was as tenuous as the use of sans-serif typefaces in the New Typography. The distinction has not yet been grasped by the wider public. The complaint is commonly heard that 'everything looks the same now': buildings are like upturned matchboxes on stilts; sideboards are the same except on a smaller scale; electric heaters are the same except that they are smaller again. There is truth in the complaint but this is attributable to lazy and pedestrian designers, not to the principles of design. In fact, so diverse is the handwriting of the true creative designers in the second half of the twentieth century, it often seems impossible that they are using the same language. What are the links between the neo-Baroque of Brasilia, the precise frame construction and curtain walls of Mies van der Rohe and his followers, the elegantly powerful use of concrete which is practised in Japan, the lovely brickwork and traditional use of timber by Alvar Aalto? What is common to chairs by Eames, Jacobsen, Race, Bertoia, Day?

It is clear that the new tradition—unlike, for example, Art Nouveau—is not a matter of style. Of what then does it consist? Four principles suggest themselves in particular as differing from those of earlier ages.

First: there is a basic assumption that no problem is so familiar, its traditional solutions so proven, that it can escape total re-assessment from every point of view. Nothing is to be taken on trust. Fresh analysis, fresh data, fresh techniques, materials or production methods, may lead to a solution of a totally new type; they may on the other hand give renewed force to solution-patterns long since established. Neither the one nor the other is to be considered better *per se*, but the design *process*—or at least the collection of data on which the first hypothesis is to be formed—for even the most familiar of objects is set in train as though that object had never previously existed. As a result, the final form of that object, if the designer is talented, has the conviction of new creation.*

Second: the requirements of man constitute the starting point for all design thought. Function was indeed the central doctrinal article that bound the pioneers of the modern movement together (though seldom proclaimed as excluding other qualities). The concept of function has come, however, over the years, to be regarded less narrowly; less mechanistically. Thus, although ergonomics and the most sophisticated data are increasingly employed at the analysis stage of design, in order to ensure the maximum degree of adaptation to the physical needs of the user, so, equally, is function taken properly to embrace his spiritual, emotional or sensuous needs. 'For all of

* One recall's Lethaby's (anonymous) note in a 1905 *Architectural Review:* 'If only we could see our building as a new problem without precedents'.

us', wrote John Voelcker of the young architects forming the nucleus of Team X, 'the individual and his needs were paramount . . . we refused to accept a polarity between individual and society.' 'The research which is above all sacred to me', Denys Lasdun has said, 'occurs between me, my colleagues and those who are going to use the building . . . if you cut the umbilical cord—the lifeblood of exchange of information with the building user—we could not operate.' The change in emphasis over thirty years is well symbolized in the person of le Corbusier himself, who could write in the early 'twenties that 'La maison est une machine à habiter'* and in the nineteen-fifties design the intensely emotive church of Notre-Dame du Haut at Ronchamp. Time was when designers required all domestic bric-a-brac to be junked as unfunctional, and designed away the mantel-piece just to make sure. Now, man's need to have about him unstandardized objects of unusual qualities is recognized no less for the ordinary adult than for the *Vorkurs* student. Scale, human scale, is a constant preoccupation of architects as buildings and urban complexes grow ever more vast. The solution may escape the designer but recognition of the problem's existence is acute. 'Everywhere we perceive the lack of a convincing humanistic perspective of technical civilization', writes Dr Maldonado.† Professor Buchanan's consideration of the functional aspects of traffic in towns led him to consideration of environmental values. It is from man's total requirements, and an appreciation that the requirements of any living organization will change, that the design brief and the design process draw their authority. In this may be traced something of the social and moral considerations proclaimed by Ruskin, Morris and Gropius.

Third: all design thinking is now informed by the concept of continuous space. For centuries solid form was seen as displacing its own volume within finite, defined spaces—in the case of a house, within the hollow cubes of its rooms. Solid and space are now felt as forming a different aspect of one joint continuum without limit. A house and its setting interpenetrate each other: floor-to-ceiling windows, patios and internal courtyards serve to dissolve the barriers between 'outside' and 'inside', so that the landscape flows into the interior spaces. Within the house open planning and split levels mean that rooms and spaces lead continuously into one another without break. Outside the house, in the planning of urban communities, the same flow of space is sought in multi-level circulation, in intricate patterns of courtyard and square, passage and underpass. Sculptors—notably Henry Moore and Barbara Hepworth—were quick to exploit this new sense of space and furniture designers were not far behind. Chairs and storage units are open, pierced, cantilevered and supported free of the ground by minimal metal *pilotis;* fireplaces—if required—have been accorded freedom to move out from the wall into the centre of the room; blinds, space-frames and open-weave

* *Vers une Architecture,* 1923. † In *Ulm,* 8/9.

materials are used to define areas without destroying the sense of space beyond.

Fourth: every problem is to be solved as elegantly as possible, using a minimum of means for the precise end in view. This may mean a design as spare as a Miesian tower, but not necessarily. What it does mean is that all elements must be fully integrated into a total design concept; that ornament and surface decoration, no less than structure, are not to be considered in isolation but only as part of the very fabric of the form; that the design must be appropriate to the material available and the means of production, and must be such as to use these without waste. Economic use of the physical means available is likely to imply the use of a modular system, all along the line from building construction to typography, from CLASP to—until recently—the covers of Penguin Books. It is perhaps here, in the rationalization of quantity production, that architecture and industrial design most obviously come together as joint inheritors of the Bauhaus philosophy. Indeed, as industrialized construction systems in ever greater numbers come through the development stage and into common use, it is increasingly tempting to see them as large-scale industrial design, rather than industrial design as a minor relation of architecture.

These are simple generalities—too diffuse, it may be thought, to be glorified as 'the new tradition'. Certainly, they represent an approach, an attitude of mind, rather than the detailed rule book of classical architecture or le Brun's codified canon. They stand at the opposite pole to that Beaux Arts training which was so rigid in its dependence upon multi-axial symmetry of plan that Ferran, for example, used to show no elevations but leave them to be inferred. This is their strength: that they have freed design from stylistic preconceptions and demand that every problem be considered in its own right, that every solution arise from the given factors. Is this then no more than a pretentious way of describing chaos? Is that ideal unity which made possible the Gothic cathedrals, Renaissance Florence, Georgian London, no nearer our grasp? The attitude of mind these principles represent is shared from Tokyo to Brasilia, Helsinki to New York. Any group of trained designers, brought together to work on a particular problem, would bring to it differing personal qualities of judgement and invention; they would not, however, argue the premises on which judgement had to be made. Not for a very long time has such a thing been possible, and never on so wide a scale.

To judge how far this new accord can work in practice, it will be necessary to study in some detail how things actually get designed and first to examine a little more closely the nature of the design process itself.

3. The Nature of Design

Nineteenth-century romanticism turned the artist into the high priest of a secularized society. Creativity is not solely the prerogative of the artist, however. Many activities are creative—for example all forms of academic and applied research, many forms of governmental, industrial and commercial administration. The creative act is in each case an ordering of existing knowledge, interlocked with an imaginative reaching out to apprehend hitherto unperceived relationships. It is seen most clearly in the fine arts because of their autonomous nature: serving no other purpose than themselves, they exemplify creativity unalloyed and without impurities (which is not to say creativity of uniform profundity). In the applied arts, in science, in the affairs of man, the picture is blurred by the given factors which have to be taken into account (which is not to say that the element of creativity is necessarily rendered less profound thereby).

The artist's creativity is concerned with aesthetics, with what is pleasing to the senses. Painters and sculptors, hand craftsmen, architects and designers share a visual and tactile sensibility and express their creativity in the solution to visual problems. Substantial differences, however, divide their activities in practice.

Art is the expression of fundamental truths that are not to be apprehended except in visual form. Art is disinterested, functionally useless, unique. The artist is a problem-solver of sorts but the problems he solves—in the twentieth century—are of his own setting. He is his own client; his is the brief; he alone decides the criteria by which the result is to be judged. It may be thought that such conditions as he may choose to set himself are more or less irrelevant, except in so far as they serve to fire his excitement and stretch his imagination to its limits. In the last analysis it is not the formal structure or technical efficiency of his work—even less its overt iconography—which is its real subject and its *raison d'être*. It is through that more fragile something which lies behind the structure and the handling that art justifies itself with intimations of eternity.

Craft is manual skill, serves a purpose, is unique within a general type pattern. The craftsman, too, to a very large degree, writes his own brief. He is less free than the painter or sculptor on two counts—his chosen materials dominate his handling of the problem to a greater degree; functional and utilitarian considerations condition its form. Nonetheless, the craftsman's approach to his task retains much of the artist's intuitive improvisation. He works to a large degree spontaneously, progressing in many cases by trial and error, primarily to please himself through domination of his materials and technique.

Design is preplanned form, useful, multiple. It is prose to the poetry of art. However much the artist and craftsman may seek to rationalize their attitudes to their work, the end product remains, in essence, an irrational or chancy affair. The dimensional relationships in even so 'mathematical' a painter as Mondriaan prove on analysis to be subjective. To say that something came about by design, on the other hand, is to say that it did not happen by chance. For the architect, the engineer, the industrial designer, design is a matter of preplanning and remote control and method. The designer must work, ice-cold, with a set of given requirements his only materials, a deadline his only spur. Not for him the clay quickening under his fingers, the ineffable moment when, almost as in a trance, the painter feels the brush taking on an independent life of its own and an image developing, as it were of its own inner necessity, on the canvas. The design process for him consists of little more than the simultaneous resolution by rational means of a number of interlocking, and mostly conflicting, problems—technical, functional, economic. Designing is analysing, planning, foreseeing, compromising, adapting means to ends, meeting other people's requirements, meeting deadlines, guaranteeing results. So mechanical is much of the process, so inevitably shaped is it by factors beyond the control of the designer himself, that the exact point where the creative spark flares, and the precise extent of the area shaped by the creative will, remain uncertain. Indeed, it is a measure of a design's success that its aesthetic, functional and material aspects are so utterly fused that they are not to be isolated. This is perhaps particularly confusing in a period when aesthetics have reached a degree of specialization only a shade less extreme than that reached by science and technology. We have become accustomed to consider aesthetic elements in isolation—separately distilled as it were—but chasing the creative element in the problem-solving activity of design is rather like trying to analyse the purely formal qualities of a late Rembrandt self-portrait without any reference to its insight, compassion and other psychological and humanist undertones and overtones.

Elusive though the role of aesthetics in the total design process may be, however, its existence—or absence—is made manifest by the end result. In this we have the evidence of our eyes. We are here, it is true, treading the dangerous quicksands of value judgements. There are no ultimates in aesthetics—tradition, fashion and our own psychology affect our view of what is pleasing. The 'mental set' with which we approach a design is conditioned by the criteria established by informed taste, but taste itself is continuously developing. Mr Bruce Archer has very exactly described the process* as 'like the case law of our courts of justice. Each judgement is based upon precedent. Each decision is added to the collection of judgements, and each added decision changes both the centre of gravity and the field of application of the collection. The next decision is thus made on criteria different from those on which the previous decision was based. In stable conditions in the course of time, a consistent but gradually adapting standard is reached.' From such schemata do we decide that this table is better looking than that, one building more beautiful than another.

The territory embraced by the word design is considerable. It is bounded on one side by pure engineering, where nearly all the factors are determined by sliderule or computer; on the other, shall we say, by book illustration, where nearly all the factors are determined by the designer alone (not all: the proportions and size of the drawing, the method of blockmaking, the number of colours, the quality of the paper; possibly the subject itself; whether it is for a right-hand page or a left-hand page; whether, in magazines, the drawing is to include superimposed lettering or panels of text—these are some of the things over which he will not normally have any control). This territory embraces one-off problems like the construction of a bridge, a ship interior, a town centre, an exhibition, all the way through to the quantity production of every kind of capital and consumer goods. It embraces work done by single designers and work done by teams of scores of specialists.

In every case the design process begins with research and analysis. The data required by the designer will vary widely from one field of practice to another but in product design, for example, some or all of the following questions will have to be answered before he can begin work:

(*a*) What is the need for the new product? This will entail market research into public requirements and demand; and analysis of the strengths and weaknesses of such existing designs as may already be available.

(*b*) What is to be the product's precise function and use? This will entail research into ergonomics or 'human engineering', safety factors, reliability of performance, and user and legal requirements of every kind (for example, car headlights must be white in Britain, must be yellow in France). Often national and international standards are so tightly drawn that they all but prohibit innovation.

(*c*) What are the problems of production? This will entail research into the qualities of different materials (possibly new and unknown), into difficulties of tooling, process and labour costs (affected, for example, by the tolerances required, capital expenditure available for plant and tooling up, the extent to which it may be possible to rationalize assembly by the use of standardized components and so on) and all the

* In his *Systematic Method for Designers*. Council of Industrial Design, 1966.

problems of how best to tailor the product to the capabilities of the plant and work force available.

(*d*) What are the problems of distribution, marketing and maintenance? This will entail research into questions of handling, packing and storage; into sales appeal and current taste (the inevitable movement of fashion is of particular importance in matters of interior design; in many fields the designer must accurately foresee shifts of taste that are likely to occur several years ahead).

The greater the number of points—social, political, economic, functional and aesthetic—at which the design problem touches society, the more complex, obviously, does their resolution become. Even the designer who is closest to the fine artist—the illustrator—can rarely hope totally to avoid this preamble of research. Mr Anthony Gross and Mr Lynton Lamb are among those who have given sensitive and illuminating pictures of the process as it applies in their own work. Of the eighteen months which Mr Gross decided he would need to illustrate *The Forsyte Saga*, six months, he decided, must be allowed for full-time documentation.

Research, then, produces a mass of factual material that is relevant in greater and lesser degree to the problem in question. This material must now be subjected to analysis. Desiderata may prove self-contradictory; some of the factors established mutually exclusive, so that prominence given to one will tend to subdue the potency of another. Often, for example, the designer will find himself demanding conflicting qualities of his materials—at one moment that they be malleable for working, at another strong for use. The rational disposition of the elements of an electronic system may fail to accord with the ergonomic disposition of its controls and instrumentation. The designer must now, therefore, by conscious decision, establish the rank order, or priorities, of his sub-problems and judge which factors are most vital to his overall purpose. He is, at this stage, rather like an aeroplane pilot beginning to move down the runway. He has completed his pre-take-off drill, has checked his instruments, is developing full power and is about to hurl himself into the moment of take-off: the moment when the rational, organizational requirements of his job fuse with the intuitive, imaginative leap into the total design concept. He has filled himself with everything there is to be known about the problem, to the point where it is pounding in his very bloodstream. His waking thoughts revolve around the problem, perhaps sleepless hours at night. He is obsessed with it. A kind of psychic pressure builds up within him, of which he can only unburden himself, only obtain release, through a satisfactory solution to the problem. In this sense, the act of creation can become almost an act of self-preservation, without which the pressures within would become so intolerable that they would lead, finally, to madness or collapse.

The moment of fusion, when form and content are suddenly, miraculously, married, has probably been preceded by a period in which the mind was prepared and relaxed by thumbnail doodling, graphic contemplation as it were, first attempts to pin scudding thoughts and images before they vanish.* Suddenly, almost by a process of automatism, the design idea or hypothesis itself takes charge and establishes its own quintessential form. At its most extreme, the designer may feel that he has been presented with the solution as by an act of fate.

If this, to continue the analogy, is the moment of take-off, the whole flight still lies ahead: scores, hundreds, perhaps thousands of decisions, adjustments, changes of course, height and speed, before the point of destination is reached and the final landing safely made. So, too, must the designer now face the longest part of the whole process. He must subject his tentative solution to the most critical examination; shortcomings will demand modifications, or even a fresh start. This second period of analysis will lead, probably, to a definitive solution—subject only to all the refinements, simplifications, elaborations and rethinking called for by the development stage of the design; and by possible changes of policy, and departures from the original brief called for by the client. All this will demand an intensely rationalizing intelligence. It is nonetheless true that all this subsequent activity must prove fruitless—in design terms—unless it finds its roots in the earlier act of creation. To put it the other way, you cannot fly an aeroplane until you have got it off the ground.

Within this general framework, however, there is room and need for further, if more limited, exercise of imagination and invention. What the designer is faced with is an area of decision that is constantly decreasing as the job progresses. Within the total design concept already established, major design decisions may well remain open; these taken, decisions of lesser importance must be faced—and so on to completion. Imagination and creative ability will thus play their part at all stages though, as the decisions decrease in importance and the problems become more routine and familiar, solutions will more and more tend to be based upon past experience and rules of thumb than upon the true imaginative leap.

So clear cut may appear the sequence of solutions presenting themselves during this development and elaboration period that the basic concept seems almost to have taken on a life of its own. Gio Ponti has described thus the design of the Pirelli building:

* Conversely le Corbusier has written: 'When a task is entrusted to me I am accustomed to place it inside my memory, that is, to allow myself to make no sketch for several months. The human head is so made that it possesses a certain independence. It is a box into which one can pour pell-mell the bits of a problem. Let it "float", "marinate", "ferment". Then, one day, out of a spontaneous burst of the inner being, the click is produced. One takes up a pencil, a piece of charcoal, a coloured crayon, and one gives birth on the paper; the idea comes forth—the child is born, comes into the world. It is born.'

Once the building had been conceived and stamped with its exact functional character, the five designers, and with them Nervi and Danusso, found themselves following its development more and more from the outside and seeing to requirements which emanated increasingly, and almost independently, from the building itself. The building, so to speak, gradually assumed its own form, its own individual entity—an increasingly precise and autonomous entity—which reported directly to the architects, with ever narrowing alternatives, such irregularities as still existed and had to be eliminated. Thus the building, after the formal-functional hypothesis that conceived it and marked it, led the designers towards its own structural invention, its own originality—nothing bizarre but faithful to its character. Thus there disappeared, 'at the request of the building', the initial central arris, so that the form might fit the structure; thus the big middle pillars came to the surface; thus their ends were elongated so that they would not appear to be flat and would be seen to be separate; thus the roofing came to be 'detached' from the structure to become an aureole. Thus the architecture almost imperceptibly blended with the structure and, without compromising but rather perfecting functional coherence, took on a formal coherence.

Where performance is paramount, or expenditure likely to be exceptionally heavy, the areas of decision may be reduced to minimal proportions by a prolonged—and expensive—period of design development. For example it has been stated* that before the Vickers-Armstrongs VC-10 took the air (and then to encounter drag difficulties necessitating further modifications), 43 major models had been built and tested in 21 wind tunnels; and that the design and building of these models had taken some 450,000 man-hours of work. Mr Ove Arup,† structural engineer for Sydney's ill-fated Opera House, has stated that: 'In the course of seven or eight years we alone (i.e. his own firm) have spent more than 375,000 man-hours on the job and over 1,800 computer hours . . . It has taken us six years to decide the final design of these shells'. The undoubted value of the mock-up is taken to extremes in the United States. A nice picture is drawn by Henry Dreyfuss of guinea-pig travellers, some of whom had packed for a one-way transatlantic crossing, others for a three-month cruise, being conducted into eight mock-up staterooms constructed in a New York stable, there to settle down for a non-existent voyage. And again of a full complement of passengers occupying an aircraft mock-up for ten hours, while lavatories, storage, facilities for serving meals and so on were tested under flight conditions. $150,000 spent thus, says Dreyfuss firmly, is worth it. Where less is at stake in material terms, the designer is rarely given comparable facilities for development of the design idea—or certainly its aesthetic elements. Though he

* *The Times*, 22 March 1963.

† Paper read to Pre-stressed Concrete Development Group, London, January 1965.

the past design development took place by trial and
or, over long periods, often centuries. The North
a herring buss, above left, is a design of 1584; her
ccessor, above right, dates from 1768. The later boat
s handier sails, and has abandoned the foremast,
ich had to be lowered every time the nets were shot.
the extent that the shapes and structure of objects
common use were refined by generations of craftsmen-designers, rather than by one man starting from scratch, they could be said to have been produced by a form of group practice.

Today the process of designing is enormously speeded up. Research, analysis, formulation, development and production are now concentrated into a fraction of the time it took to evolve the elegantly functional Berkshire wagon, below.

—G.P.D.

ANTHROPOMETRIC DATA — ADULT MALE SEATED IN VEHICLE

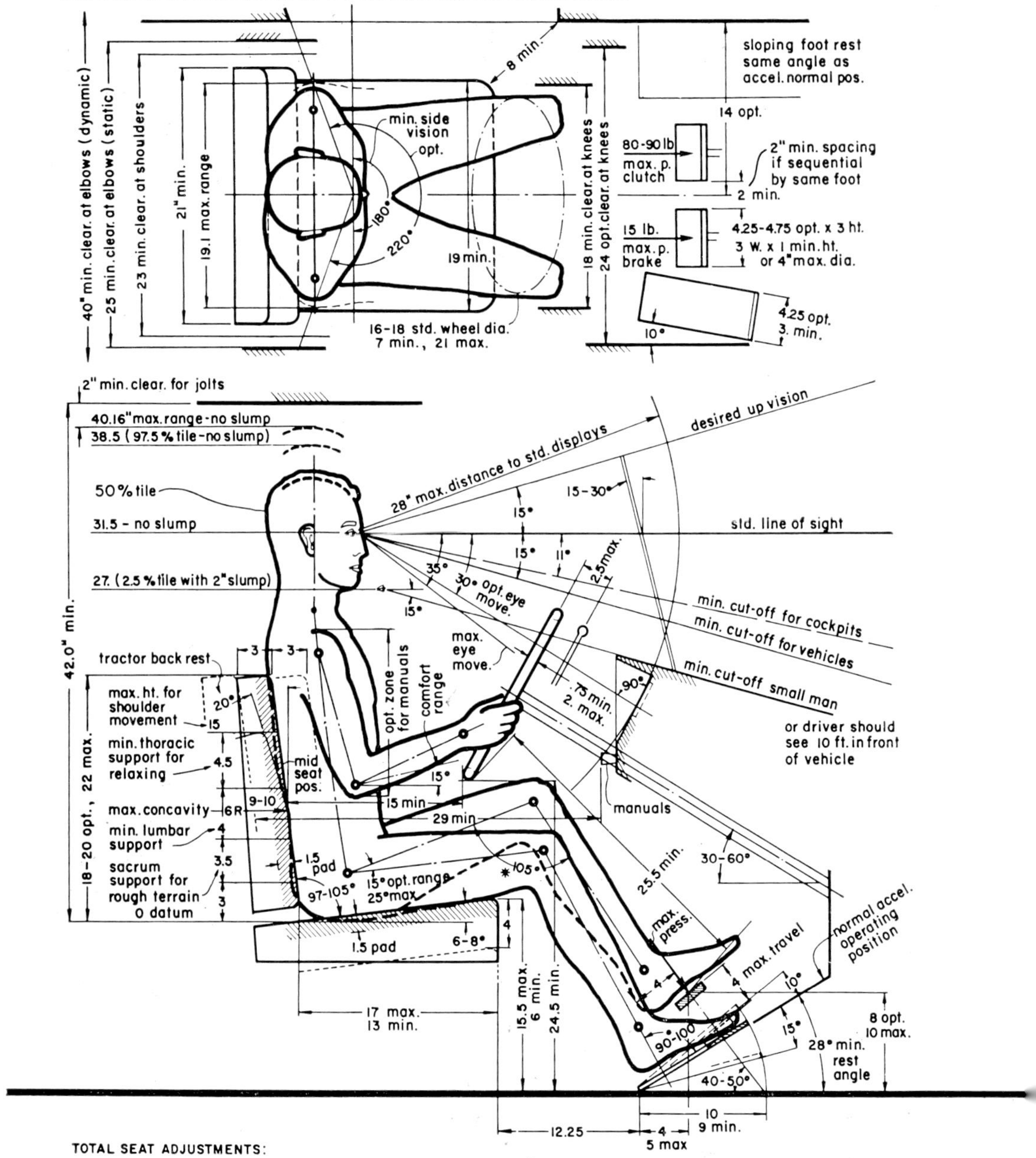

TOTAL SEAT ADJUSTMENTS:

horizontal: 6" min. in max. increments of 1"
vertical: 4" min. in max. increments of 1"

* leg angle 105-110° for max. pedal pressure 0-50 lb.
120° min. " " " 50-100 lb.

WHITNEY PUBLICATIONS N.Y.C.

Design requires data *and* the imaginative leap. To meet today's parameters, an ever greater volume of information is necessary if the designer is to do his job. Ergonomic, stress and fatigue, and wind-resistance tests are only some of the research operations that may have to be tackled. For ergonomic tests, adjustable r are commonly constructed; but before that stage is reached the designer will need such anthropometric data as is given in this page from Henry Dreyfuss's *Measure of Man*.

rmed with all available
formation, and after appro-
iate analysis, the designer
akes the imaginative leap
wards his solution. Sym-
lic of this indefinable
ocess is Paxton's giant
ctoria regia (now *amazonia*),
e underside of the leaf of
nich suggested to him the
uctural system of his
ass-houses. Another such
mp was made by Issigonis
en he suddenly saw the
plications of placing the
gine transversely in the
MC Mini.

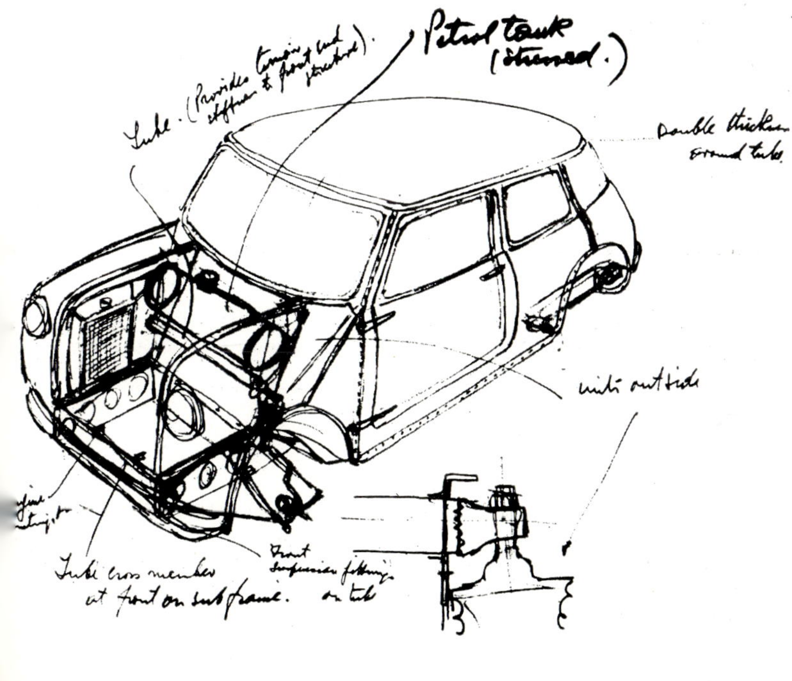

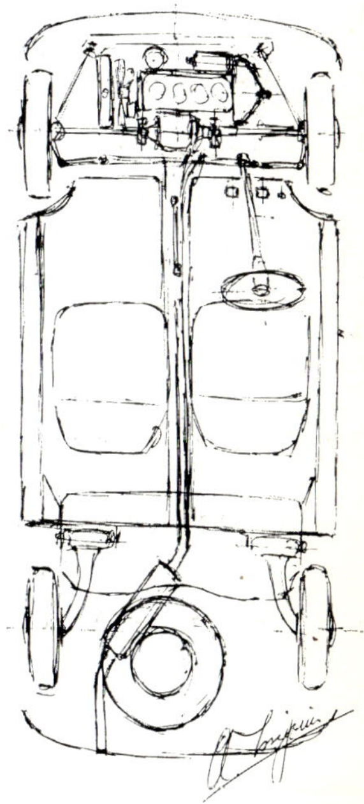

The purpose of design is to bring conflicting factors within a common framework or conception. From this act of synthesis, this resolution of warring claims, the structure or object or image emerges with a new clarity. On these pages are two examples of the results of industrial design in action. Above, the Gallic lathe, in production until the mid-sixties; below, its successor redesigned by R. G. Tallon.

Opposite, the redesign of a communications system, essential purpose of which is to ensure the easy and rapid assimilation of multiple information: signal boxe St Pancras, above, and Birmingham New Street, bel

In a complex world the unit of design tends to grow ever larger, embracing not merely individual objects but whole ranges of objects. A 'corporate identity programme' will bring buildings, products, printed matter and all other aspects of an organization into a common design framework. Shown here are different uses of the symbol devised for British Rail as part of such a programme by Design Research Unit, partner in charge, Milner Gray. The application of the programme is controlled by a detailed instruction manual, above left—the twentieth-century equivalent of the eighteenth-century pattern books.

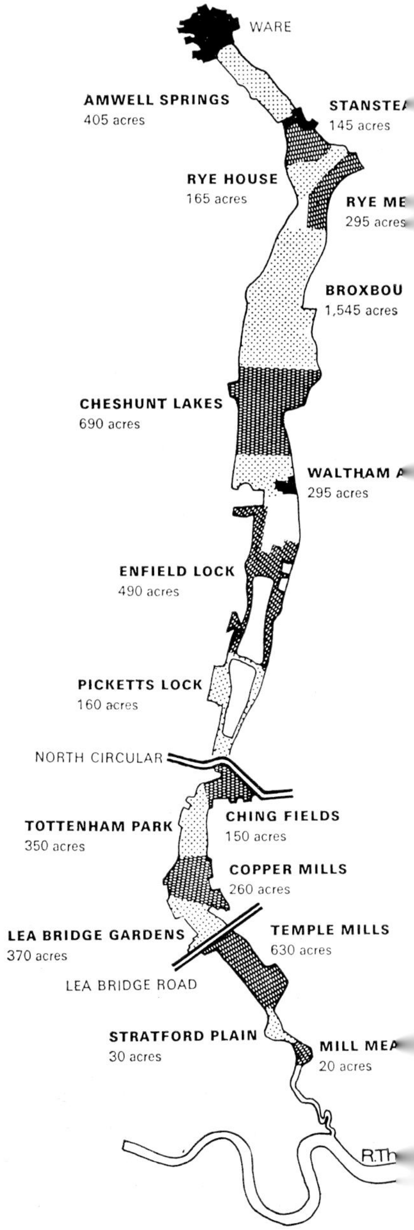

Environmental design has to bring factors many times more complex and numerous into a common relationship—on an urban, regional and even national scale. Right, the Lea Valley Regional Park, now the responsibility of a special Authority set up by Parliament as a result of proposals by the Civic Trust. The Park will be more than 20 miles in length: it will link and unify land previously divided between eighteen local authorities.

·early such large-scale undertakings are beyond the ·pacity of the individual designer; the team is born. fferent skills form new patterns of collaboration—on · *ad hoc* basis for specific jobs, in continuing associa-·n, or permanently in an integrated office. Above, one the mid-century's happiest examples of architect-painter collaboration: the architects—Skidmore, Owings and Merrill, a very large group practice; the painter, Miró. Below, a very different kind of artist–architect teamwork: plan for housing at Peterlee New Town by the painter Victor Pasmore in association with the Corporation's architects.

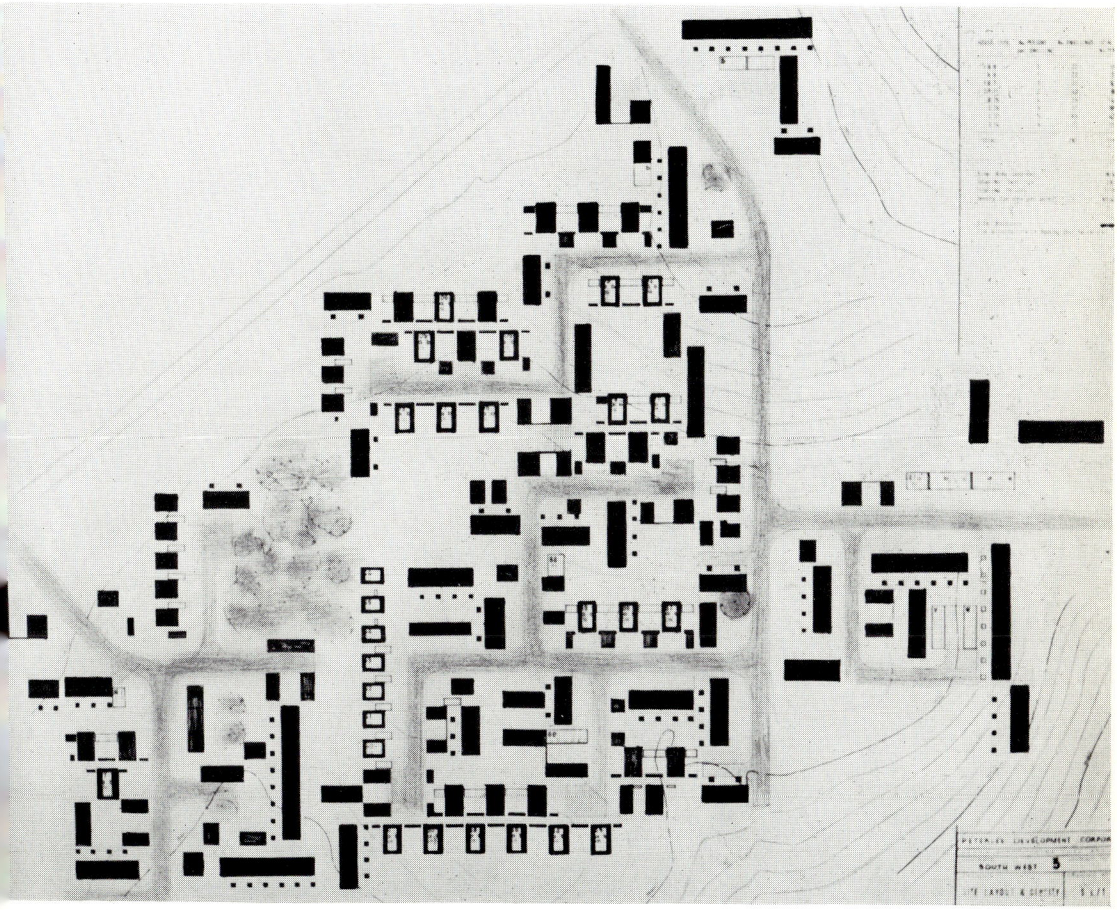

Buildings by multi-disciplinary practices in Britain. Above, Graduate House, Leckhampton, by Arup Associates, architects and engineers, for Corpus Christi College, Cambridge, Below, barracks at Aldershot, by Building Design Partnership.

may be applying his own personal and arbitrary tests for excellence of colour, tone, form, texture, line and the rest, rather than sifting the data arising from scientific and technological testing, the *process* remains similar and it is often the designer's wish that he could prolong it. He may well be forced by deadlines and delivery dates, however, to complete the design before he is fully satisfied with it.

Pressure of time may warp the design process in a multitude of ways. Sir Gerald Barry has recalled how, owing to the tightness of the timetable and in the absence of final decision by the Government on the site to be used, the theme of what was to become the South Bank Exhibition in the 1951 Festival of Britain had to be devised without knowing the exhibition's size or whereabouts. Ranelagh, Hurlingham, 'somewhere in the City of London', Battersea Park, the exhibition halls of Earls Court and Olympia, and the national museums were all under consideration. 'We had to start', said Sir Gerald, 'by producing a thematic plan without any physical plan into which to fit it—a plan, that is, capable of execution on any scale, in any place, according to the ultimate choice of site and design of buildings. This was a little like the Lion's advice to Alice about the cake, "Hand it round first and cut it afterwards".'*

Accidents happen. The west wall of the Erechtheion, on the Acropolis, had to be withdrawn to the east after building had started, lest it impinge on the adjoining tomb of Cecrops. The north porch, that is to say, was in all probability not part of the original design idea but a 'saving operation' in substitution for an intended portico to the west. It is by no means uncommon for the designer to find his very terms of reference being amended at an advanced stage of the whole process and not always because of indecision on the part of his client. The South Bank Exhibition of 1951 can again provide an example. Seventeen architectural offices were invited to submit designs for seventeen different buildings, an essential part of the brief being that they should utilize new materials and novel methods of construction, so as to spare the traditional materials—which were then still in short supply in Britain—for housing. The architects fulfilled their brief all too well. When the designs came in some months later, the Festival office was warned by its consulting engineers that the buildings could not be completed in the time available—owing to the novelty of their construction. Only two were accepted; the remainder had to be scrapped or, in some measure, revised.

The essence of successful design is to bring the object or structure concerned into satisfactory relationship with the entire system of which it is part. This is often rendered impossible by social, administrative or economic factors quite beyond the control of the designer. Henry Dreyfuss records the case of an aircraft interior in which the varying seating densities demanded

* Second of three Cantor Lectures, delivered at the Royal Society of Arts in May 1952.

by the operators made it inevitable that some of the seating patterns could not be made to accord visually with the fixed window pattern. Again: the chaotic disposition of street furniture, cumulatively and increasingly a major factor in the destruction of the urban scene, is an inevitable result of divided responsibilities. It is notorious that the lanterns, columns and sometimes brackets of our street lights in Britain are separately made by different groups of manufacturers without reference to each other; that none of these manufacturers has working contact with the makers of the litter bins which will in many cases be strapped to the columns. Even were it possible to persuade firms making the main items of street furniture—bus shelters, seats, guard rails, signs, bollards, traffic lights, parking meters and the rest—to adopt a common modular system, it would still prove impossible to group the results in an orderly way while a multiplicity of public authorities and utilities retain the right to erect their own structures without reference to one another.

To retain his creative grip on the situation in circumstances like these calls for resilience, toughness and patience on the part of the designer. He must acclimatize himself to the knowledge that success in design can rarely, if ever, be absolute. The most that can normally be claimed is that, the limiting factors being so and so, within the context of what was possible the end result provided an apt and elegant solution. What every designer dreams of, and rarely gets, is the brief adequate enough to make even this possible. Sir William Holford has put the point with clarity. 'Design', he said,* 'is no longer seen as likely to emerge by a sort of process of selection or personal taste from a more or less standardized emotional vocabulary of architectural parts of speech. Design is now coming, I think, to be recognized generally as a skilful answer to a brief, and so it stands to reason that imagination and method have both got to go into the writing of the brief itself.' The task, in fact, 'demands a degree of objectivity and analytical experience which is a design exercise in itself'.†

If the creative faculty of the designer is to operate at full stretch, the brief must present a challenge. Once the past has been shed as a main formative influence, it is only by looking forward that ideas can be generated. Man's highest artistic achievements have always broken new ground. They have resulted from their authors' sense of pitting themselves against problems that have previously seemed to defy solution: sometimes material, as in the Van Eycks' use of oil colour or the Cubists' use of *collage;* sometimes technical, as with the discovery of perspective or the laws of light; sometimes emotive, as in the Mannerists' deliberate distortions or the Surrealists' encounters outside time and space. It may be very modest or very ambitious: a belief that

* 'The Role of the Architect/Planner', *Architectural Association Journal,* 4 December 1962.

† Jack Howe, writing in the *Financial Times,* 27 April 1964.

no one else has ever really caught the quintessential character of a potato, or tunnelled right through a block of stone before, or worked on such a large, or small, scale. Whether or no this is truly so in fact is irrelevant. It is the artist's *belief* that he is contriving to make a statement that has never been made before which lends to his endeavours a special cutting edge and which may, at its highest, actually extend the syntax of aesthetics. The excitement it engenders informs the whole work and can still communicate itself across the centuries long after other attributes the work may have had contemporaneously have grown musty or incomprehensible. Conversely, the conscious knowledge that he is adding *nothing* to the tradition in which he works can but depress the temperature of what the artist is doing, to the point where life leaves it.

In the crafts different considerations apply. Function and materials equally restrict expression. Man's physical requirements have not changed greatly through the ages: pitchers have held water, cups have held tea; glass has been blown, wood carved, silver hammered, wool woven. It would be an arrogant craftsman who claimed for himself a greater skill than the makers of Tutankhamen's throne or the Chinese bronze-casters of the seventh and eighth centuries, the mosaicists of Ravenna or the scribes and illuminators who produced the Book of Kells.

Architecture, engineering and design can perhaps be placed, in this respect, somewhere between the arts and the crafts. Although shaped largely by function, materials and tradition, they are capable nonetheless, when the designer has felt himself to be working at the limits of human experience, at the frontiers of knowledge and his times, of glories that can move the spirit. At such moments the fusion of form and function has been complete.

The nineteenth-century engineers were such initiators. Operating in new fields, without precedents, they were forced to create their own terms of reference, their constructional laws, their technical means, *ab initio,* working only on logical premises. If a railway company required a viaduct or a bridge of a certain span, the only means of discovering whether the brickwork would bear such and such a load, or whether the tolerances of the ironwork would permit the span to meet in the middle, was to construct it and see. If the solution to the main problem entailed the solution at the same time of sub-problems—for example the use of a new method or material—that too had to be evolved *ab initio* and by the same designer. The thrust and breadth of his imaginative thinking at once solved, and arose from, the challenge as a whole. He saw it whole because he had to.

Some of those confident Victorians scarce gave themselves time to learn to read or write. How, one wonders, would they have fared in the twentieth century? Today the very size of projects, the scale of financial investment involved, the technical complexity of construction and manufacture, militate against any one designer achieving this kind of total synthesis unaided. The present century has seen so vertiginous an extension of human knowledge that in science, for example, it is now said to double in volume every ten or fifteen years. Whole areas, previously capable of comprehension, are tending to pass beyond the mental grasp of the individual. For the designer, no less than the scientist, the volume of data relating to functional requirements and techniques on the one hand, the forbidding extension of uncharted territories in the fields of materials, production and construction on the other, together make increasing specialization inevitable. From this in turn two broad tendencies spring: the one leading to the development of systematic design techniques; the other to the need, for any but the simplest projects, to create multi-disciplinary teams to replace the individual designer.

The employment of linear programming, critical path studies and logical decision structures, of anthropometric data, of cybernetics, of computer techniques and matrix charts, of consumer research studies and similar aids, would be over-complex for many problems but can be invaluable in certain types of large-scale ones. Such methods are not without their own difficulties and dangers, however. Three may be instanced as examples.

Industrial design has a particular concern with ergonomics, or the man-product relationship—'human engineering' as the Americans call it. Ergonomic studies scored their first successes in transforming the exceptionally bad cockpit layout of early military aircraft, the discomfort of which, with the opportunities for error they presented, resulted in avoidable strain for the pilot. Such studies are now routine, in some measure, for most structures and equipment for human use. The application of anthropometric data, however, even when obtainable, is by no means easy and, unless handled with experience, can lead to results less satisfactory than those achieved by the designer's intuitive understanding of the problem. Thus, a dimensional range appropriate to, say 80 per cent of potential users may be considered adequate in relation to one characteristic; if, however, as is probable, other characteristics have to be catered for in addition, the 80 per cent coverage will fall sharply, for only a percentage of the original 80 per cent will fall within the range set for these secondary and subsequent characteristics. Philip Johnson's view of the ergonomic chair has already been quoted.

A second danger may be detected in the weight given to consumer research and the extent to which public preference—or what is thought to be public preference—may come to dominate design policy. This is one of the great pressures of our day. For example, the Political Correspondent of *The Times*, reviewing a study of the 1964 General Election in Britain, wrote thus:*

> The real risk is that we are moving towards the day when market research, opinion poll findings, techniques of motivational persuasion and public relations, and even the analyses of political scientists will be crudely and

* *The Times*, 22 April 1965.

> cold-bloodedly used to govern party strategies in government and out. When that day comes—sometimes it already seems to be here—the point of politics will not be the conviction of politicians about the rightness of their principles, policies or beliefs. It will be to discover the degree of prejudice or gullibility in the electors. The point of leadership will not be to lead, but to follow the crowd.

Too often the radical, the fresh, the unusual design—even the worthy pedestrian design—has been jettisoned by a management in favour of some vulgarity it believes the public will prefer. Such craven attempts to curry favour with the customer will not become respectable simply because they are decked out with pseudo-scientific trimmings. In the view of IBM's consultant Director of Design, Eliot Noyes, consumer surveys have 'led to all kinds of corruption in designers and in their designs'. Even more briskly, the designer's position has been put by Denys Lasdun:*

> Our job is to give the client, on time and on cost, *not* what he wants but what he never dreamed he wanted and, when he gets it, he recognizes as something he wanted all the time.

A third danger from such methods could perhaps affect the designer himself. Increasingly, the design process is split into specialist fragments, each component problem being solved so completely in technical terms that no element of doubt, of choice, remains. At no stage is there any question that the problem can and will be solved as a matter of routine. The emotional temperature falls, the imaginative leap becomes a little hop, the designer has no longer felt and experienced the work through in all aspects himself. The likelihood is that the end product will consist of an acceptably linked series of acceptably solved problems: that it will work precisely as predicted but that it will not be a unified work of art, as Telford's Menai bridge or the Villa Malcontenta are works of art.

There are those who deny totally the aesthetic element in design—or at least in design that is not concerned with surface decoration. In this view the final form and appearance of a product can properly arise only from its precise matching to function and construction; if these receive optimum consideration, there is no room left, it is argued, for aesthetic manoeuvring. This rigorous doctrine can be traced back to Hannes Meyer,† who succeeded Gropius at the Bauhaus in its last years. Max Bill, a Bauhaus student at that period, was in turn to create the Hochschule für Gestaltung at Ulm, now under Dr Maldonado as Rector, where methodology and systematic techniques are paramount. Ulm has gone so far as to dispense with the graphic design training that normally forms part of design school curricula, and has

* In a talk at the Royal Institute of British Architects, 9 February 1965.

† Gropius has since written, in a letter to the *Architectural Review*, March 1963: 'I believe that Hannes Meyer's inner downfall was his denying art as such. It narrows the field, if the rational point of view is made the only factor.'

substituted for it the study of 'communications'. In point of fact, designs of great elegance have emanated from Ulm—best known in Britain being the chaste electrical equipment manufactured by Braun of Frankfurt. The truth of the matter is that there are few products in which the functional and structural demands are so strict that there is no room for aesthetic decisions in their design. Aircraft are popularly believed to be of such a kind, any beauty they possess being a by-product of inevitable form. Misha Black, however, has recorded his surprise at the extent of the area remaining open to modification on aesthetic grounds, when he was called in as consultant designer on a light 'executive' aircraft.*

The real purpose, then, of the application to design of sophisticated methodologies is not to supplant the creative function of the designer but to help him to avoid wasteful dispersal of his creative gifts—on the wrong problems at the wrong time, on chasing blind-alley hypotheses, through confusion between performance desiderata and aesthetic desiderata.

A distinction must of course be made between those fields of design in which performance is an important, if not a paramount, factor and those, more decorative, in which it plays a minor part. The former are susceptible to computer aids and methodology; the latter are not. There is a considerable area of overlap between them at the perimeter but it is nonetheless clear that, for example, most forms of graphic design, of interior design and of craft-based design—textiles, glass, ceramics, silverware, jewellery—are and always will be shaped primarily by human sensibility.

At the same time all types of design, including these latter, are increasingly tending to be brought within single comprehensive design briefs—and rightly so. Particularly is this evident at either end of the design scale. Not only, at one extreme, is serious physical planning likely to require land economists and sociologists, geographers and ecologists, surveyors of various kinds, town planners, traffic engineers, civil engineers, architects and landscape architects. At the other extreme the directional sign on the station platform, the menu card in the restaurant car, the typography of the time-table, the numbering on the locomotive, are seen, not as separate problems, but as aspects of one overall problem which can only be solved satisfactorily as a whole. The scale of expertise required by such operations, from the creation of an environmental complex to the 'corporate identity' programme of commerce, is inevitably causing the design function to pass from the grasp of the individual into the hands of teams of specialists and is reflected in the impetus towards larger design offices which is everywhere apparent. Whether such teams are in fact permanent, or whether they consist of individuals and small practices brought together on a temporary basis, within a properly co-ordinated framework, for a particular purpose, does not greatly affect the nature of the operation. In either case it can be seen as the main-

* By reason, of course, of the vast surplus of power now available for small planes.

tenance of productivity through the division of labour by specialization of function. The question then arises as to how far the essentially private nature of the creative act can in fact be shared among the members of a group.

John Madge* defines the bureaucratic system normally implied by such specialization as requiring 'a clear social organization, with an explicit hierarchy and a rigid chain of command and discipline'. He quotes the German sociologist Max Weber: bureaucracy, the latter considered, is characterized by its 'precision, speed, unambiguity, knowledge of the files, continuity, unity, strict subordination, reduction of function and of material and of personal costs'.

All forms of design, as has been seen, involve much routine work. In engineering, in architecture, in graphic design, this is normally done by the draughtsman, the junior, the assistant. Such juniors, that is to say, are employed non-creatively within just such a system as that described by Weber—most noticeably where the head of the firm is himself a designer of great and idiosyncratic talent, and in those great offices which are more concerned with the efficient and businesslike handling of large contracts than with the quality of their design.

Is there no alternative? At variance with this conception is set the idea of the group team, composed of talents that are inevitably various and unequal, but which are given the fullest opportunity at every stage to make to the project as a whole such contribution as they may be capable of. In the fullest sense—not easily achieved—the essential purpose of group practice is to link and focus the creative and critical faculties of every member of the team, not just upon one or two facets of the problem but upon every aspect at every stage.

On all sides, those concerned with problem solving can be seen to be groping towards new patterns of collaboration. Indeed, the very attention being given to methodology, resource allocation and greater productivity in design offices—itself an aspect of increasing specialization—leads naturally to renewed consideration of management problems; of the presentation of information for decision; of the relationships between superiors and subordinates, among the professions and between the professions, their clients, contractors and manufacturers.

This perhaps is the central problem of design today and it demands examination at greater length.

* In a paper prepared for the RIBA's Management Advisory Group and printed in the RIBA *Journal* for November 1964.

4. Patterns of Collaboration

It has long been convenient for professional people to form associations that enable them to share their experience, their overheads and, it may be, to even out the work load upon their staffs. In the traditional partnership, however, each partner handles a separate job, retains his personal professional relationship with the client and may not feel the need to seek even informal advice about the work in hand from his partners. Within their common discipline, partners may well, indeed usually, have special aptitudes or experience which enable them to complement each other's abilities, thus enabling the practice as a whole to cover a wider range of activities than would be possible for any one of its partners individually. Traditionally, however, the professional disciplines did not intermingle within an office. A firm of architects requiring the services of a quantity surveyor or a structural engineer would employ other firms of chartered surveyors or civil engineers as consultants on the specific project concerned. A firm of civil engineers constructing, say, a hydro-electric scheme might employ a landscape architect as consultant to advise on how best to integrate the scheme with its setting.

Today such relationships begin to appear unbearably cumbersome and, more often than not, downright inefficient. It is true that many professional firms, through working with each other repeatedly, have established close working relations. It remains evident, nonetheless, that the process of decision-making must be greatly slowed by the need to ship documents backwards and forwards between offices, if staff have continually to journey afield for conferences and progress meetings, and if information is offered by specialists without full knowledge of all the factors involved.

We are here at the heart of the matter. Not merely do traditional methods of consultation militate against efficient communication; they militate against specialists making their full contribution to the solution of a problem. For example, the quantity surveyor is today moving into a much more positive role in the building industry. He is developing methods of cost control which directly affect the design process, indeed form an essential part of that process. Both quantity surveyor and constructional engineer, for example, provided they are built into the team at an early enough stage, may be able to propose solutions that will permit a different allocation of the budget within the agreed total to the advantage of the job as a whole: perhaps an alternative structural system that will free more money for finishes, or a saving on the total operation that will benefit the client. This is not possible where they are simply costing or calculating a building already fully designed by the architect. It is the perpetual frustration of the designer, be he landscape architect or typographer or product designer, that he is called in too late, when all major decisions have been taken and the project has already assumed such a form that little can be done to it save to clean up some of its superficial ugliness. This is not design. The elegant design solution is that which meets the maximum requirements with the minimum means. This postulates that *all* relevant factors must be embraced by the creative act of synthesis which we call design. This, in turn, means that if the factors are so complex and so numerous that a number of specialists are necessary to comprehend them, then *all* those specialists must be involved, on equal terms, at each stage of decision-making from the very beginning of the process to the very end.

There is wide acceptance of the necessity for new patterns of inter-professional collaboration, for multi-disciplinary teams. Practice lags behind theory, however. In Britain, over the last ten years, the lead has been taken by the larger units of the construction industry. You require a factory, an office block? We will find the land, negotiate and purchase it, obtain planning consent, design the building, construct and landscape it, finance it. The process, for busy clients exasperated by the multiplicity of fragmented services they must otherwise call upon, is streamlined and painless; its attraction obvious. Not dissimilar is the service offered by today's advertising agency, whose package deal can embrace market research, sales policy, choice of media, copywriting and design, space-buying, film production and a fistful of other specialized activities.

From the standpoint of the professional, such teams tend to be suspect because their final control is vested in other than professional hands and the client relationship is not therefore necessarily governed by the professional's own ethically based code of conduct. Professionalism, it has been said, implies an unwritten contract between the professional and society, over and above his duties to his client. Professional advice is as objective as intensive training and a long-tried system of checks and sanctions can make it. It is uninfluenced by commercial, political or other pressures. It is offered to the client only at the request of the client, and in so far as he may wish to keep it so, it remains an entirely confidential exchange between professional and client. Take away this relationship and the client—and society—are left only with such safeguards as may obtain in normal commercial practice. This is not a situation with which the senior professions, in Britain at any rate, can be associated.

At the same time the building and design professions cannot escape those same pressures which have brought about the grouping of functions represented by the 'package deal': the impetus towards larger units and greater productivity, the need to rationalize administrative procedures and to offer the public simpler points of reference in a complex world. A report to the Minister of Public Building and Works by Sir Harold Emmerson in 1962* stated:

> The fact must be recognized that the private architect operating in many small units does not meet modern requirements in such matters as organizing ability, cost control and office management. More businesslike methods are needed to meet the challenge from the comprehensive design and building service offered by many building contractors. Improvements could be made if private practitioners engaged in similar practices were to join forces, sharing experience as well as overheads. Also a way must be found of establishing closer contact with other members of the building team.

Additionally, many in these professions—though by no means all—retain the dream of an ideal unity between all those creatively involved in a problem, a unity which could liberate power of a new order. Here, for instance, are two architect-planners on the inter-professional group. Professor Colin Buchanan: '. . . most important of all is the need for a new synthesis of minds among those professionally engaged in these problems. A much improved concept of professional collaboration is needed, with an emphasis on mixed team working.'† And Mr Graeme Shankland: 'I would suggest that many of the more important problems of society can only be tackled in this way. I am convinced that not only can the job only be done in this way but that, if it is, you can move a mountain.'

A mountain of inertia and prejudice will certainly have to be moved before the multi-disciplinary team becomes general. In Britain there is, among the professions, a good deal of polite suspicion and much gentlemanly jockeying for position. It is not always even clear where the boundaries should be drawn between them. Engineering, over the past half-century, has continued to bifurcate with the energy of a multiplying cell, so that there are now 30‡ separate institutions and societies dealing with civil, marine, electrical, lighting, municipal, sanitary and a score of other branches of engineering. Only now, with the formation by the thirteen chartered institutions of a Joint Council and the receipt of a Royal Charter, can the first movements towards re-unification be detected. The planning profession is deeply split between those who see it as a meeting point for all the disciplines concerned with physical planning: geographers, economists, sociologists, architects, surveyors, landscape architects, lawyers and the rest; and those who see it as an autonomous discipline in its own right. The very title of the Society of Industrial Artists and Designers reflects a certain strain between those of its members concerned with two-dimensional design and those concerned with three dimensional design.

* *Survey of Problems Before the Construction Industries.* HMSO, 1962.

† *Traffic in Towns.* HMSO, 1964. ‡ Listed in Whitaker.

If some of the professional bodies are not entirely at one with themselves, how much greater are the gulfs between them. Their Codes differ—for example, civil engineers can and do operate as contractors, which is expressly forbidden to the other professions. Fee structures differ. Educational syllabuses differ and the very use of language often differs. Some are large: the Institution of Mechanical Engineers embraces some 55,000 members, the Royal Institute of British Architects about 20,000. Some are small: The Town Planning Institute has around 4,000 members, the Society of Industrial Artists and Designers about the same number, including student members, the Institute of Landscape Architects a mere 350 Fellows, Associates, Licentiates and Student Members. Such differences do not reflect the importance of the contribution made by practitioners in these fields; neither do they make for balance in the debate on inter-professional collaboration.

For recent years have, indeed, seen discussion on these matters removed for the first time from the theoretical level to the institutional. Many of the professional bodies concerned with design and environment are now committed officially to movement in the general direction of greater integration. Thus the Noel Hall Committee, set up jointly by the RIBA, the RICS, the Institute of Builders and the Institution of Structural Engineers, in presenting its report early in 1965, set out a minimum common syllabus for six areas of interest in the building industry. Several British universities and schools have anticipated such reports. The Bartlett School of Architecture has a common syllabus for the first two years of training for architecture, environmental engineering, building, landscape design and planning. Sheffield and Nottingham are among those pioneering a more closely integrated pattern of building education. The disastrous split between engineering and industrial design, emphasized by the Feilden Committee, has been shown to be far from immutable by the Royal College of Art. At another level, the Association of Consulting Engineers recently amended its rules to make it easier for members to enter multi-professional practices. Such developments, striking as they may be in the context of past history, are but the opening shots of what must prove a long campaign. At workaday level the sort of difficulties that will remain until traditional prejudices are cleared away was strikingly shown by the RIBA *Guide to Group Practice and Consortia*.* Several cases were there instanced of firms and individuals who, though in fact practising in association, nonetheless felt unable to publish the fact, except to their clients, through fear of losing business in their own firms. Thus 'Consortium F1', an amalgamation of two architectural offices with

> . . . a wide range of specialist advisers—structural and service engineers, quantity surveyors, planners and landscape architects . . . met with considerable opposition from other professional colleagues in the locality, who suggested that the arrangement was unethical. The services

* RIBA, 1965.

consultants concerned received a number of protests from other architect clients, threatening not to give them further work, because of the feeling that their allegiance would be to the consortium rather than to their other clients.

In the freer atmosphere of the United States, the barriers between the professions have always been much less rigid. For example, the office of Walter Dorwin Teague, himself originally a typographer, regularly undertook engineering quality control work. Architects, engineers and designers agree that paramount importance attaches to the formation of a design team at the start of a complicated project and lose less time in profitless debate as to who should lead the team (it being clear that this must vary with the problem).

In America the work of architects, consultants and contractors is much more closely defined, agreed and understood than in the United Kingdom. This leads to better relations between the professions and greater efficiency (though the added freedom and responsibilities of the American contractor can, on occasion, lead to loss of control over design by the architects and engineers).

A distinction is made between the 'team approach', in which both architects and engineers collaborate, usually under the 'chairmanship' of an architect, in the initial conception of a scheme, and the more traditional approach in which the architect conceives the scheme, taking advice from consultants as he requires it.

In practice, the collaboration takes two forms. In Chicago and Boston it is customary for architects and engineers to work together in a single firm of architectural and engineering consultants. In these cases, the architects usually employ the engineers (although the reverse was also noted). In New York, it is usual to find separate firms of architects and of engineers combining, as in the United Kingdom, to execute a particular contract; and both engineers and architects seem satisfied that early and proper collaboration is possible with this arrangement.*

Working within this freer framework that lacks some of the checks and controls of British practice, architects and designers are less inhibited by professional squeamishness from moving into areas outside those of basic design services. As with the big contractors, this is partly to meet the needs of their clients, partly through recognition of the fact that 'in many cases, if it does not participate in the early real estate and financial operation, and similar decisions, the architectural firm will be prevented, by decisions over which it had no control or influence, from applying its best skills and talents to the solution of the more strictly architectural problems'.†

* 'Building Research—current papers.' Design Series No. 6. Reprinted from the RIBA *Journal*, 1963.

† Dudley Hunt, Jnr, 'Comprehensive Architectural Practice for Small Offices or Large'. AIA *Journal*, September 1962.

Growing trends are discernible, the same writer continues, 'towards consultation between individual architects and firms, associations and joint ventures of various kinds'. He goes on to suggest that 'an architectural firm with a reputation for original and enlightened design might acquire a commission for which they intend to handle only schematics, preliminaries and over-all co-ordination. The aid of knowledgeable real estate and financial consultants is enlisted for assembly of the necessary land for the site, analysis of financial requirements and obtaining of financing. A specialized consultant is called in to prepare an analysis of the operational and building requirements and eventually to prepare the programme under the direction of the architects. An association is formed with an efficient and business-like firm of architects who are given the task of production of working drawings and specifications. Since the building is to be constructed in a location away from the home offices of either firm, still a third architectural firm of demonstrated ability in the area of construction is engaged to manage and administer the construction of the building....'

Allowing for the differences in the nature of the job, industrial design in the United States appears to be moving in the same direction. The barriers within America's swiftly developing industry, indeed, have never needed breaking down as they have in Europe. When Norman Bel Geddes took on Standard Gas Equipment's range of stoves way back in 1932—and he refused to do so except on his own terms—the first thing he did was to install one of his own engineers in the Company's plant. Comparative studies were made of competitive products. Field reports of hundreds of interviews with dealers and housewives ran to over 300 pages. The happy result is part of design history: hundreds of separate parts were replaced by 16 standard components and sales were doubled.

If this was still exceptional in the early nineteen-thirties, American industry embraced the industrial designer totally after the war. As a result the designer has tended to broaden the service he offers. A forceful firm may undertake research and feasibility studies of all kinds, ergonomics, product naming and marketing, store planning and interiors, corporate identity programmes and other specialized activities. For instance Raymond Loewy's office, now Raymond Loewy/William Snaith Inc., undertook for the American Meat Institute a study of 'how meat should be butchered, packaged, displayed, sold, advertised, labelled, cooked and served'.* For Montreal's Cultural Centre they determined what activities should be embraced by the Centre's programme and how those activities should be financed. Norman Maclaren, of the National Film Board of Canada, was sent to China by UNESCO to advise on communications by filmstrip. Russel Wright, of the 1933 spun aluminium cocktail shaker, was despatched by the U.S. Mutual Security

* *Design*, 183, March 1964: a special issue on America prepared and edited by Ralph Caplan. I am indebted to the same issue for other information in these paragraphs.

Program to help determine and develop the economic potential of underdeveloped countries in South-East Asia.

The width and variety of services offered by architects and designers has become a distinctive feature of the contemporary American scene. In regard to the latter it has been described, at its best, as industrial statesmanship; at its worst, opportunism.* Certainly both architects and designers overstep the British concept of professionalism. A number of design offices employ an out-and-out sales force and at least one has its own Director of Public Relations with supporting staff. When all is said and done, however, it remains clear that on both sides of the Atlantic swiftly moving currents are eroding older patterns of collaboration among the design professions. They may be expected to flood wide areas—of pretty parched territory—during the next decades. It should be of interest to glance briefly at some of those who have already begun to explore and chart this territory.

The concept of group practice is not new among the professions. Accountants, doctors and even, up to a point, lawyers have all adopted it in some measure. With general practitioners it has become as urgent a challenge as with designers, and for not very dissimilar reasons. Among architects and designers, the idea of semi-anonymous collaboration, in which the shared experience and aspirations of the group would supersede the romantic concept of self-expression, seems to have been born in the early nineteen-thirties—the fruit equally of an increasingly functional design idiom and an idealism which perhaps reflected the radicalism of the decade.

Between 1933 and the outbreak of war a number of such groups were formed, most noticeably in Britain and the United States; more followed with the return of peace. Not all have survived; some have flourished dramatically. Of those that are still in existence, most found it necessary to adapt their original structure over the years to meet circumstances and conditions that were not foreseen at the outset. Indeed the actual form of collaboration adopted by these firms differs widely in detail. Collectively, however, they shared and share a largely common attitude of mind, a common approach to the job—what has been called, in a wider context, a 'diversity of talents with a unity of values'.† Among some of the better-known examples, the following may be instanced briefly (some to recur in greater detail in later chapters).

The year 1933 saw the formation in England of *Tecton* and the *Industrial Design Partnership*. Both, it is interesting to note, involved men of Russian birth. The former was the creation of the dynamic Berthold Lubetkin, who came to England via America. The latter marked the association, among others, of Milner Gray, Misha Black and Kenneth Bayes. Tecton embraced

* Ralph Caplan, Ibid.

† Professor C. A. Mace, *The Psychology of Study*.

some of the liveliest young architects of the day on a basis of democratic equality: the partners shared the work and fees, each partner taking on those jobs for which he was most suited. Among the work produced was that brilliant modern classic, the Penguin Pool at the London Zoo and, outstanding in its day and still impressive, the Highpoint blocks of flats in Highgate (1935). The group was inevitably thrown out of balance by the impact of Lubetkin's personality and began to break up, in its original form, even before the outbreak of war. It continued in existence under Lubetkin, nonetheless, until 1948 and members of the original partnership have on occasion worked in association since then (e.g. the Holford Square housing estate in London by Skinner, Bailey and Lubetkin, 1954).

Industrial Design Partnership, itself the successor to an earlier group known as Bassett Gray, ceased operations with the coming of war but, as is described later, reconstituted itself in more ambitious form in 1943 and continues today as Design Research Unit, one of the largest groups of its kind in Europe devoted primarily to industrial design.

The significant partnership of Louis Skidmore and Nathaniel Owings was effected in Chicago in 1936. Not until 1949 was the group structure of the by then vastly increased, firm established in its present form, but the seeds were present from the outset. Only a year after the partnership was formed, Skidmore, in answer to a commission stipulating a partner in residence, left for New York—never to return. Decentralization had begun. From the beginning, too, Skidmore and Owings had decided that the firm must embrace all the skills. Severinghaus, Cutter and Brown brought experience in housing, hospital design and industrialized building techniques respectively. Bunshaft's exceptional design abilities were to establish SOM's standards of quality. John C. Merrill, a structural engineer, was brought in as a 'limited partner' in 1939. In Skidmore's words,

> Up to the war, most U.S. architects were trained to work only on small plots. They left the problems of coping with large-scale projects—industrial plants, airfields—to the engineers. We felt that if the architect wasn't to limit himself to domestic housing, he would have to win back his role as the creator and co-ordinator of big projects.*

The history of the firm since then has become one of the wonders of the architectural world.

Next may be instanced the *Architects Co-operative Partnership*, now *Architects Co-Partnership*, born in 1939 of the association of eleven ex-students of the Architectural Association School who had left the year before. All worked in different firms but each put up £10 and with this capital a basement office was set up. From here, by means of evening work after their normal day, the practice began. In the nature of things, it was only after the war that the firm really got going as a whole-time practice. Their

* Quoted in 'The Architects from "Skid's Row"', *Fortune*, January 1958.

rubber factory at Brynmawr, begun in 1945 by the partners without assistants, marked a notable breakthrough in British factory design, by reason of its then exceptional integration of architectural form and process requirements. Today, with seven of the original partners still in the firm, ACP has become one of the most considerable and distinguished of British practices, responsible for a very wide range of work in Britain and Africa. It comprises, in addition to the partners, six associates—mostly acting as group leaders and looking after a number of jobs, between 75 and 100 technical staff and 20 administrative staff. Every job is the responsibility of a Partner; Anthony Cox has said that in ACP Partners, as well as being responsible to the client, have a special sense of responsibility to each other also. Partners meet on alternate Mondays to discuss policy; on other Mondays with Associates to discuss work, design and techniques in relation to material presented by a job architect. Group architects meet fortnightly to exchange views and monthly staff meetings are held for more general discussions of current work or extra-mural activities.

About the same time that ACP was setting up its basement office, three one-man architectural practices in Preston were forming themselves into a group to carry out important war work, merging to all intents and purposes into a single firm two years later. The moving spirit here was George Grenfell Baines, whose initiative the firm recognized in its name: the *Grenfell Baines Group*. In 1944 they took on three more architects, a quantity surveyor and a secretary; a further architect joined two years later. 'These members shared the income of the firm (fees less all expenses) on a basis of mutual assessment; each member evaluated the percentage worth of his own work contribution and that of his colleagues, the final shares being determined by averaging the assessments.'* By this time, however, individual ambitions were conflicting with group loyalties and the scheme—it seems that no legal partnership agreement ever existed—was terminated by mutual consent. The Group split into several traditional practices, four of which nonetheless remained in loose association and continued to use the old title as a subsidiary to their own main names. In 1961 the architectural firm of Grenfell Baines and Hargreaves, already with five offices, in London and the north, joined forces with the quantity surveying firm of Arnold Towler, with the object of forming a comprehensive design group. As is described later in greater detail,† the result of the marriage, *Building Design Partnership*, a firm comprising by 1966 eight offices in different parts of the country, has shown a sustained determination to embrace all the skills required by the building industry, from engineers to industrial designers, within a balanced and democratic structure.

* *Guide to Group Practice and Consortia*. See also Chapter 12, section 1, for Building Design Partnership's current practice.

† Chapter 12, Section 1.

Mention should next be made of *Arcon*, a somewhat specialized collaborative practice brought into being in England after the war, with the object of advising industry how best to channel its wartime production facilities into the manufacture of components for industrialized building systems. The three partners each brought a specialized experience to the problem and the firm's success was its undoing, since the commercial use of the name (the Arcon school, etc.) was to prove an embarrassment professionally and the practice did not endure.

In the United States, the end of hostilities saw the formation, in January 1946, of *The Architects' Collaborative*. Gropius had asked one of his ex-students, John Harkness, to assist him at the Harvard School of Design. Simultaneously four young graduates from Yale got in touch with Harkness about the possibility of setting up a group practice. A spur-of-the-moment decision to join forces received equally quick encouragement when both first and second prizes in a competition were received by four of the new partners. The finances of the firm were at first ordered on an almost Marxist basis of need, each partner's basic drawing account being supplemented by an increment for each child (an arrangement discontinued after financial success made it unnecessary). Sarah Harkness wrote, in 1957,* of the office as follows:

> The first year consisted of very few jobs, but a great deal of discussion. The system of job captains and office meetings was set up. There is a meeting every Thursday noon to discuss either design or business. Meetings are called in between times whenever a subject needs special attention. Every job has to go through a meeting before the design is frozen. The job captain, together with those who have been assisting him, brings drawings, models and pertinent facts and figures to the conference room. A really 'hot' design meeting often produces the best architecture. In the end the changes are those of the job captain himself, because he has the final authority.

Today TAC has fourteen names on its letterhead and behind it such buildings as the Harvard Graduate Centre, for which Gropius was job architect and which was awarded the 1951 Gold Medal of the Architectural League of New York.

Three more post-war examples may be cited. The *Architect's Department of the London County Council* exhibited for the last twenty years of its life, between 1945 and 1965, many of the characteristics associated with group practices. The Department was at once not only multi-disciplinary, embracing all the environmental skills from town planning to furniture design, but was organized on a group basis of well-nigh autonomous teams, averaging perhaps fifteen in number. That this great organization, some 3,000 strong, was created by Sir Robert Matthew and his successor Sir Leslie

* *Architectural Review*, May 1957.

Martin within the framework of English local government makes it no less an example of group practice—save only in legalistic terms—than if it had been a private firm. Its works are legion: the Roehampton Estate, the Crystal Palace Sports Centre and the ill-fated study for Hook New Town are but three of the first to come to mind. History may find its greatest legacy was none of these. Through the Department passed, in those two decades, a very large proportion of the brightest and best of the country's younger architects, planners and designers, to take with them in their subsequent careers something of the LCC's standards and the LCC's concept of teamwork.

Italy has seen the growth of the *Studio Ponti-Fornaroli-Rosselli* in Milan, a remarkable circus ring in which architecture, industrial design, arts, crafts and publishing have been put through their paces as part and parcel of a single performance. Chief impresario is Gio Ponti himself, who not only turns his hand with brilliance and dash to architecture, furniture design, ceramics, textiles, to industrial products such as coffee machines and to editorial work, but has shown an equal flair for bringing outside painters and designers into the team whenever it has seemed appropriate. Though he himself set up in practice in 1925, it was really when he joined forces with the engineer Antonio Fornaroli in 1930 that the essential inter-disciplinary basis of the firm was established. The interests of the two older partners are now well complemented by those of Alberto Rosselli, architect and industrial designer with a particular concern for methodology in the design process. The Pirelli building (1959), designed with Nervi and Arturo Danusso as structural advisers, will remain an enduring monument to the Studio's architecture; the monthly *Domus* to its concern with design and publishing.

A recent addition to multi-discipline teams has been *Arup Associates*, formed in 1963. The name of Ove Arup, born in England of Danish parents and himself educated and trained as an engineer in Copenhagen, is inseparable from the development of British architecture since the early 'thirties. He collaborated with *Tecton* throughout, joined the MARS (Modern Architectural Research) Group and in 1938 started his own firm of engineers and contractors with the object of more closely integrating design and construction. He withdrew from the firm at the end of the war and since 1946 has given all his time to consultancy, almost entirely devoting himself to advising architects on the calculation of structure and its integration with architecture —with what success may be gauged by the growth of his firm, now numbering all but one thousand, and the receipt of the RIBA's Gold Medal for his services to architecture. It seemed an appropriate and logical step, therefore, when Arup Associates was set up to provide for a permanent association of architects and engineers. The total staff, after three years, numbered over 70, of whom 25 were architects, 15 were structural engineers, 15 services engineers, 10 quantity surveyors, 3 furniture designers and 3 model-makers. The practice operates in five groups, each embracing the main disciplines within the firm; three groups are led by architects, two by engineers. Arup has said: 'In a team working for a common purpose in a friendly atmosphere, it can happen that something emerges which is more than the sum of individual contributions. In Arup Associates we are trying to eliminate the psychological barriers to teamwork, to create a kind of composite brain for each job.'*

Honoured names these, but they do not begin to form a comprehensive list. The LCC was indeed pre-eminent among local authorities but it was not unique in its desire to create balanced teams. For example Coventry has gone further in setting up integrated teams for particular projects drawn from the City Architect's and Engineer's Departments. Almost any consultancy practising town and country planning must, practically by definition, constitute a multi-disciplinary group. Other countries, it may be assumed, are no less conscious of these problems and anxious to meet them adequately. The Philips Architectural and Engineering Bureau at Eindhoven, it is said, has found it necessary to group its 225 architects and civil, mechanical and electrical engineers in integrated design teams. In a different sense, something of Gropius' mantle seems to have fallen upon the modest shoulders of Kenzo Tange in Japan. Within his office at the top of one of Tokyo's highest buildings, Tange leads a small team of around twenty young architects. Work is developed by joint discussion and criticism among the group. Notwithstanding his quietness of manner, Tange is said to have exceptional drive and ability as a team leader and draws out the abilities of others with sympathy and skill. Though ornament as such is rejected, Tange does not hesitate to co-opt artists into the team as and when appropriate. The relative anonymity of group practice must commend itself, it may be thought, to the oriental mind and it would not be surprising if the example of this manner of working were to spread.

It will be clear from these examples that the tag 'group practice' has long been used loosely to cover one or more of a whole cluster of ideas. At different times it has meant quite different things to different people. It has been applied to various forms of association between independent firms within a single profession and between independent firms in different professions. It has been applied to single practices in which different skills and disciplines are associated. It has been applied to single firms which practice a large measure of devolution, either geographically among their various offices or in the creation of semi-autonomous teams within one office. It has been used of offices organized democratically rather than hierarchically, in which decisions are reached through discussion by the *group* rather than passed down through a chain of command. In any particular case these things may be found singly or in combination.

A clarification of terms is clearly desirable and this the Royal Institute of British Architects has largely done for us in its valuable *Guide to Group*

* Address given on the occasion of his receipt of the RIBA Gold Medal, June 1966.

Practice and Consortia. The *Guide* defined the association of architectural firms for mutual convenience and to render an improved service as 'group practice'. The grouping of firms of different professional skills it defined as 'consortia'. The bringing of different professional skills into one firm at all levels including partner level, it defined as 'integrated practice'.

Within each of the first two categories it identified five different forms, or degrees, of association. These ranged from loose schemes of association for pooling information, through the increasing sharing of facilities—library, accountancy and so on—through the co-ordination of other practices for a specific project, the sharing of staff, job responsibility and finance, to short- and long-term forms of joint partnership. It may well be, as the *Guide* itself states, that further forms will emerge. Clearly, however, any significant breakthrough towards the re-establishment of a living design tradition that can inform and shape the whole human environment is likely to come only from the so-called integrated practice, in further association or not with additional firms. Greater importance, it may be thought, attaches to a brave attempt to harness painters, sculptors, designers, glass workers and metal workers, scribes and all the rest, to such a common end as Coventry Cathedral, whatever weaknesses may be apparent in the end result, than in the mere linking of similar practices, however convenient and profitable that may prove to those immediately concerned.

How then do things actually get designed? To attempt to chronicle every type of design task and every kind of design collaboration would be impossibly tedious. The case histories in the next section must be taken only as glimpses—glimpses of the sort of problems the designer is today asked to solve; glimpses of his attempts to meet these in as rounded a way as the present structure of our society allows. They are not presented as outstanding successes. Deliberately, the spice of topicality and novelty has been avoided. These examples, indeed, are not intended to 'prove' anything—only to show that *this* is how things get designed; that man-made things do not just happen but that many people have worked together to shape them thus and thus.

Administratively the organizations described have little in common. They would not all of them fit neatly into the RIBA's classification of group practice. What they do share is an emphasis 'on individual freedom of initiative instead of authoritative direction by a boss'; a belief that 'sychronizing all individual effort by a continuous give and take of its members, a team can raise its integrated work to higher potentials than the sum of the work of just so many individuals'.*

* Walter Gropius, writing of TAC, *Architectural Review,* September 1953.

5. Communication

With the world-wide spread of literacy and the mass media, with the growing complexity of everyday life, with the fantastic development of communications in the plural, communication in the singular has become a matter of the utmost importance to the modern world: communication between nations, between rulers and ruled, between specialist and non-specialist, between manufacturer and market: the list could be extended indefinitely. In relation to every nook and cranny of our society, from periodicals to traffic signs, packaging to instrument gauges, instruction manuals to political campaigns, the design of communication is studied as never before. Printed word and graphic image, film and recorded sound, model and three-dimensional construction, light, odour and a multitude of other aids to perception are pressed into service as may be appropriate, to clarify and amplify the stream of signals necessary to the functioning of the human ant heap. The record that follows is concerned with one particular means of communication that calls upon many of these techniques and many different types of creative skill: the thematic exhibition.

Belief in the didactic value of a display of contemporary objects, as opposed to the historicism of museums, took root in the mid-nineteenth century. After the success of the 1851 Exhibition there was simply no holding the idea: Great Exhibitions, Empire Exhibitions, Trade Fairs, State Fairs and World Fairs, followed one another, and still do notwithstanding a chronic inability to balance their budgets (at least in the direct sense; increased trade and visitors bring an indirect reward) with inexorable regularity. All too often, however, in the modern world, increasing concern with their commercial potential has led them, notwithstanding a cultural corner for prestige purposes, to partake equally of holiday camp and sales pitch. Their framework varies from the sophisticatedly imaginative to the vulgarly tawdry—their only common factor perhaps being the visual and ideological chaos resulting from a large number of unco-ordinated organizations competing for attention.

As a purposeful medium of communication, the exhibition is of much more recent date. Only during and immediately after the Second World War was it brought, in Great Britain, to a pitch of high efficiency. This is the story of how the largest of all such projects was mounted; one which, in some measure, affected the thinking of an entire nation. To work on it was created what was probably the biggest design team ever assembled in our time. Architects, painters, sculptors, landscape architects, interior designers, exhibition designers, product and engineering designers, typographers, illustrators, modelmakers, photographers, lighting experts, hand craftsmen, structural engineers and scores of other design specialists combined to focus their creative skills in concert upon building a dream, an environment, a little utopia, that lasted for 150 days. This was the Festival of Britain.

There have been larger exhibitions in terms of acreage and expenditure. There was never before, and so far only since at Lausanne in 1964, a comparable attempt to harness the creative effort involved in an undertaking of this magnitude upon a *wholly co-ordinated result.* The Festival of Britain was not a Trade Fair, nor yet an assembly of national pavilions. It was conceived as a unity, designed as a unity, executed as a unity. Its purpose was to tell the story of a nation: that nation's 'contribution to civilization, past, present and future, in the arts, in science and technology and in industrial design'.

Here was an undertaking to tax the skill of a generation no less intensely than one of le Brun's more elaborate projects. Though ephemeral, it was widespread geographically, all-embracing in its interests and of considerable administrative complexity. Without precedents for guidance, it was required to break new ground in every direction. It had to be planned and executed in three years. Since then, the thematic exhibition as a means of communication has tended—in Britain anyway—to fall into desuetude, killed possibly by the sense of consummation engendered by 1951 itself. The Festival, however, constituted such an intensive example of the design process in operation, indeed of group practice at work on an altogether unusual scale, that it is still of interest to study the machinery that grew up uniquely to bring this exceptional project to completion.

The origins of the Festival may be traced back to a proposal made to His Majesty King George VI by the Royal Society of Arts in 1943. At the end of the war, the threads were picked up on several sides, notably by Gerald Barry, as he then was, editor of the *News Chronicle* newspaper. The Board of Trade had for some time been considering something of the sort and at the instigation of Sir Stafford Cripps, the Board's then President, a Committee was set up in September 1946 to examine the possibility. Six months later the Committee reported itself strongly in favour of a first category international exhibition to be held in London at the earliest practicable date. The Government then appointed an inter-departmental committee to consider possible sites, but in the light of access and communications difficulties posed by the 300-acre site suggested, felt bound to jettison the whole concept of an *international* exhibition. After March 1947 therefore, responsibility for this still formless idea was taken over by Mr Herbert Morrison, the Lord President of the Council.

In the autumn of that year, it was still undecided whether the project would require new machinery to run it or whether it could be administered through existing agencies: in December the Council of Industrial Design, the Arts Council and the British Film Institute were named as sponsoring bodies for specific contributions to what had now been titled the 'Festival of Britain'. Morrison next appointed a further committee, the Great Exhibition Centenary Official Committee, which was composed of officials of all government departments likely to be involved. This remained in being until the Festival came to fruition and it was indeed this committee which contrived to produce, in the words of Sir Gerald Barry, 'a skeleton for a Festival which is still recognizable today as the remote ancestor of the more sophisticated creature which, by natural selection, was to evolve during the next three years'.

By the spring of 1948, however, a special Executive Committee was created to do the actual planning of the Festival, being responsible to a Council of some thirty leading men and women from various walks of public life under General Lord Ismay. The Council was to meet five or six times a year, and was to prove to have a will of its own. The Executive Committee was composed in the first instance of representatives of the Arts Council, the COID, and the Central Office of Information (the Governmental information agency), under Gerald Barry who had been appointed Director-General at the end of March.

With the establishment at this point of the traditional executive-legislature framework, the first stage of the Festival's administrative life may be said to have been concluded. From now on the machinery was developed organically and deliberately.

The Executive Committee was enlarged by stages, eventually to include Directors for science, architecture, finance and public relations, with co-opted members for books and films. It never exceeded eleven in all. Sir Gerald Barry has recorded their initial alarm at the complexity and immensity of the task before them.

> I therefore proposed that at the earliest possible occasion we should all spend a continuous week-end in one another's company, in order to become better acquainted with one another and with the job. I could not be sure how the suggestion would be received by my colleagues; after all, I was inviting them to do something well outside their contractual obligations. But I felt that unless we could quickly succeed in establishing a strong team spirit and become united in a common enthusiasm, we should fail to fulfil our assignment. We spent the first week-end of that first May at my house in West Sussex and occupied the time in almost continuous session, beginning to hammer out the shape of things to come. Weather and company were in their best form. In the daytime we worked on the terrace or paced the lawn in pairs like walrus and carpenter, trying to marshal the oddly-assorted shoes and ships and sealing-wax at our

> command into some sort of order and coherence. Below us rolled away in succeeding folds of green and gold a landscape of English parkland inherited from an age of men who had such faith in the future of their country, and so robust a sense of responsibility to their successors, that they planted not for themselves but for their great-grandchildren. Here was our incentive. This successful week-end was to be followed by others of a similar kind. The team spirit was born.
>
> We broke up in a state of possibly somewhat exalted enthusiasm but of remarkable unanimity and determination; and thereafter for the next three or four years ordinary working hours were forgotten. The Festival of Britain was to be made not only round committee tables—though there were all too many of these for our comfort. It was made on hill-tops, in gardens, round a log fire, wherever half a dozen people could forgather and talk. It was made clambering among rubble and cement mixers, amid the uproar of cranes and pile-drivers, in overheated railway carriages and under-heated motor-cars, tearing through the English landscape. It was made in mayoral parlours, on fogbound airfields, in dingy studios, on visits to experimental building stations, in lecture halls, youth centres and standing on street corners waiting for a bus.*

As the immensity of the panorama being unrolled by the Executive became apparent, specialist guidance became essential. The Directors of Architecture and of Science were given additional support by the setting up of special advisory bodies: the Council for Architecture, Town Planning and Building Research (consisting of eight names under the Chairmanship of H. V. Lobb, plus Casson as Director of Architecture) and the Council for Science and Technology (consisting of ten names under the Chairmanship of Sir Alan Barlow, plus Cox as Director of Science). The latter was further reinforced by ten scientists who joined Cox's whole-time staff, each to take charge of one facet of the scientific story. Eventually no fewer than eighteen further voluntary advisory panels were created to advise on and check the content of the Festival for accuracy and integrity.

The whole web of history and discovery, work and play, people and things and ideas, had, however, to be integrated and interpreted in physical form. One of the Executive's very first decisions had been to create a 'presentation panel' to advise it 'on the composition, design and content of all Festival exhibitions and on the visual aspect of all Festival projects'. The presentation panel consisted of ten members—of whom three only were not architects or designers—plus a Secretary. The weight given to its work was made manifest by the Chairmanship of the Director-General himself. This might be likened to the managing director of a large commercial concern taking the

* First of three Cantor Lectures by Sir Gerald Barry at the Royal Society of Arts, 12 May 1952.

Ration books, building licences, travel allowances. Willowherb over the bombed sites. Buildings grimy and shabby. Austerity had still not been banished but the Festival of Britain had the promise of a rainbow after the storm. For five months its symbol, right, hung in the sky. Across the river the South Bank sparkled by day and glittered by night. People danced in the open on the Fairway. It was their first post-war experience of wholly designed environment.

Over the Dutch barn, and the sculptural landscape of a reclining figure by Henry Moore, the sun and moon revolve—a poetic extension of the farmer's wind pump first proposed. Below, the Countryside pavilion may be seen immediately to the right of the Dome of Discover In all, $8\frac{1}{2}$ million people visited the South Bank: and this was but one of eleven exhibitions handled directly by the Festival organization. Numerous other displays were mounted by individual bodies and, during the same summer, well over 2,000 towns and villages organized events and programmes of many kinds.

chair of the firm's production committee—with the difference that the Festival panel were producing design rather than consumer goods. At a later stage, the panel formed a 'design group'—of Misha Black, Hugh Casson, James Gardner, James Holland and Ralph Tubbs—to assume detailed design control of all Festival projects. The Group's primary task was to produce a master plan for the central showpiece of the Festival but, in addition to their shared Group function, each member came, as will be seen, to accept specific responsibility for some major area of the whole. The presentation panel and the Design Group thus became the vital, central crucible through which every creative idea relating to the fabric of the Festival had to pass.

In June 1948 the Executive considered a thematic plan for what was known as the 'combined' exhibition—the language was still guarded—the main sequences of which were, in fact, to be finally adopted in the South Bank layout. A circular chart split the theme into two 'hemispheres', that on the right dealing with the Land of Britain and that on the left dealing with the People of Britain. Concentric rings linked past and present achievements in the various fields grouped within these hemispheres. There followed a master plan which was adopted by the Executive in August and by the Council in September.

That autumn a decision was finally taken to utilize the South Bank for this main exhibition—one of the first sketch layouts (embracing, it may be noted, a large circular building on the site to be occupied by the Dome of Discovery) is dated 26 August 1948. Thereafter the abstract development of the concept, in the form of schedules of accommodation for the different sections and so on, merged into a succession of increasingly tightly organized site layouts. By January 1949 the essential framework, with its upstream and downstream areas (corresponding to the two hemispheres of the earlier thematic chart) were firmly established. The following month a Diagrammatic Materials Plan appeared, showing the percentages of aluminium, steel, timber, canvas, glass, etc.—building materials were still in short supply—to be used in the still notional buildings. By the spring of 1949, that is to say, there existed an administrative core to organize the Festival, a theme, an outline site layout and a partial schedule of costs and quantities. It now became possible to request seventeen architects, or architectural offices, to submit sketch designs, each for a particular building.

The Ministry of Works being unable to take charge of construction, the Festival Office assumed responsibility and appointed a leading firm of consulting engineers, Messrs Freeman, Fox and Partners, to take charge. As has already been noted, however, it became apparent when the architects' schemes began to come in during July that the complexity and novelty of their construction, taken with the congestion on site arising from its concurrent demolition, levelling, draining, etc., were such as to prohibit their

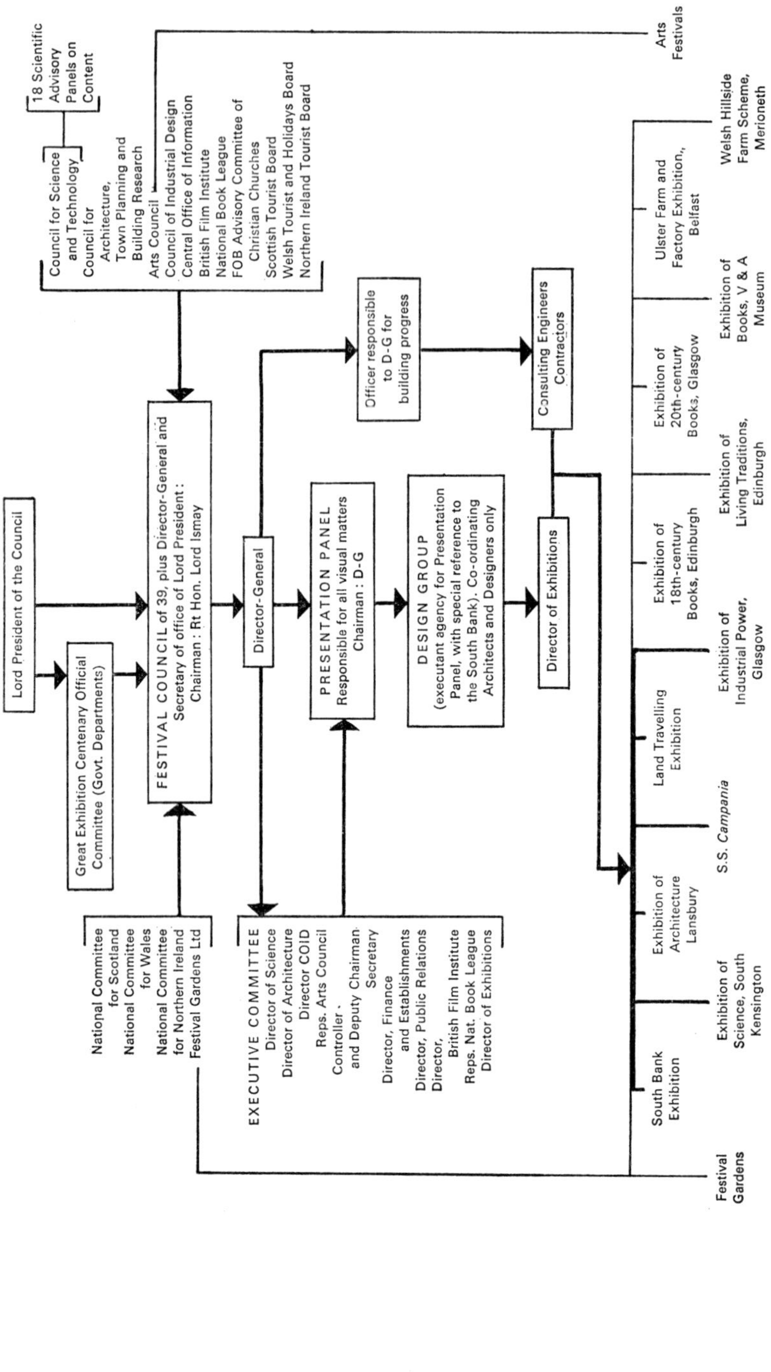

Festival of Britain: general organization.

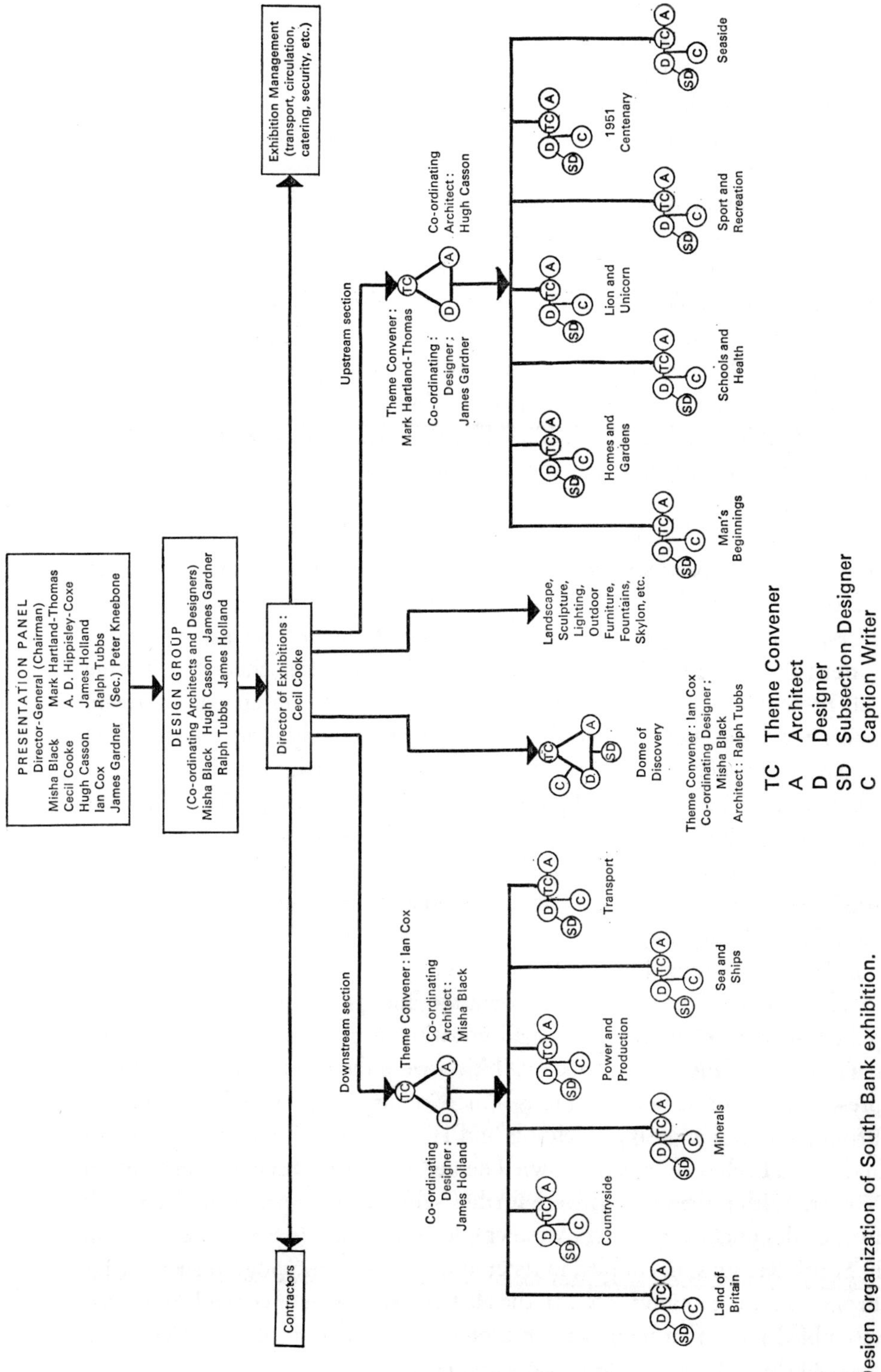

Design organization of South Bank exhibition.

completion by May 1951. Of the plans submitted only two were accepted without modification; several had to be scrapped completely.

The difficulties of the times had their own effect upon design. 'We began with the understanding that there would be virtually no steel available; only to find, half-way through, that steel became decontrolled. Softwoods, on the other hand, grew ever more scarce and in the midst of our planning we had to cut our original estimates for this material by more than half.'* Any design concept is intimately bound up with the materials in which it is expressed; the extent of the re-thinking occasioned by such changes may be imagined.

By the spring of 1950 Barry realized that the Exhibition would not be finished on time unless something drastic were done. He therefore obtained the Council's approval to the creation of a new post which would entail sole command of building progress (though not design), responsible only to Barry himself and without committees to slow things down. The Chairman of the Architectural Council, himself a member of the Festival Council, accepted the challenge. This was the last major change to the administrative structure.

If, however, the basic structure of the organism seemed adequate to its tasks, the organism itself continued to expand and develop over two-and-a-half years. Its cells multiplied steadily, all the time following the same essential pattern that gave the Festival organization its particular character. Theme and presentation were coupled at every stage. Upstream and downstream areas, the Land and the People of Britain respectively, had each a Chief Theme Convener, Ian Cox for the former, Mark Hartland-Thomas for the latter, balanced in each case by a Co-ordinating Architect, Misha Black for the former, Hugh Casson for the latter, and a Co-ordinating Designer, James Holland for the former, James Gardner for the latter. From this administrative pattern Ralph Tubbs's 'Dome of Discovery' in the upstream section, the most complex single pavilion, was excepted and was put under the design co-ordination of Misha Black.

This pattern repeated itself all down the line, as the overall problem was pulled apart into increasingly small constituent elements. Within the two main halves of the South Bank Exhibition were the twenty-two sections represented by the various pavilions—the Countryside, Minerals, Power and Production, Sea and Ships, Schools and Health, The Lion and the Unicorn and so on. Each section had its own Theme Convener, Designer and Caption Writer. Within the sections were further self-contained problems for which further designers and writers were employed. Thus the Science Exhibition in South Kensington coupled in its credits the caption typographer with Dr Bronowski the caption writer. In the abstract, the chain of control somewhat resembled that of an inter-Service exercise, with officers of equivalent rank

* Second Cantor Lecture by Sir Gerald Barry, 19 May 1952.

from the two or three Services forming joint staffs at every level. In practice the organization was much less rigid, with the same names, as we have seen, popping up at a number of different points. For example, Casson as Director of Architecture sat upon the Executive Committee, but also on the advisory Architectural Council; on the all-important Presentation Panel and the Design Group concerned with the South Bank, and was himself responsible, as has been said, for the detailed co-ordination of the downstream section of the latter. This type of multiple responsibility was common among the twenty or thirty key figures that formed the creative spearhead of the Festival organization and was clearly a factor in knitting the programme and the various teams together.

Indeed, had it not been so there is every likelihood that difficulties of agreement between the levels of the control chain, taken in conjunction with the constructional and materials problem, would have utterly destroyed the timetable. As it was, much of the design work had to be completed in four months—not very long when precedents were often lacking and imponderables all too many. The South Bank 'progress display schedule' set 12 columns against the twenty-two sections. After listing the Chief Theme Convener, the Section Theme Convener, the Co-ordinating Designer and the Section Designer for each section, it continued:

Section Designer Briefed in terms of general content (Responsibility Theme Convener)

Section Designer Commissioned (Responsibility Finance)

Preliminary Briefing—Breakdown into subsections, story sequence, essential major exhibits and main display features (Responsibility Co-ordinating Designer and Theme Convener)

Preliminary Designs from Section Designers

Preliminary Designs Approved (Responsibility Director-General and Presentation Panel)

Section and Sub-Section Designers Briefed in Detail (Responsibility as before)

Sub-Section Designers Commissioned

Final Designs, Working Drawings and Specification Completed

The first date in the schedule (for briefing the Section Designers) was 1 May 1949; preliminary designs were due about six weeks later; working drawings were due at various dates between 5 September 1949 and the end of the year.

This, it must be remembered, was only *one*, albeit the main, exhibition of eleven being handled directly by the Festival organization. Others included a science exhibition, an exhibition of 'live' architecture and planning, one of Industrial Power, the 'Land Travelling Exhibition' and the SS *Campania* which took the Festival round the coast of Britain. Beyond these there were exhibitions beyond number on every conceivable subject, arts festivals in every part of the country, activities in every town and village in the land. Administratively, this fantastic marshalling of forces, this channelling of specialist and technical knowledge, this co-ordination of design, was a triumph. To a degree never achieved before or since in a great exhibition on this scale, the schedule was maintained, and to all intents and purposes the Festival opened its gates on the appointed day complete.

In retrospect, the quality of its design now has an unfashionable, dated look. How could it be otherwise, for exhibitions are by their nature ephemeral. It has been said, truly enough, that the Festival's architects and designers were getting out of their systems all the ideas that they had been unable to put into execution during the previous decade—that the South Bank was a culmination of the 'thirties rather than a forerunner of the 'seventies. Had Britain not gone to war, had rationing not been in force and essential materials in short supply, had the site not had to be cleared of slum property and much of it reclaimed from the River Thames, above all had the designers been given more than a few months to define and complete their concepts, doubtless the Festival would have been different.

As it was, it did communicate to the general public for the first time in Britain the possibility of creating a new kind of environment—gay, exciting to move about in, full of new shapes, full of controlled surprises. It was Britain's first experience of modern architecture—the 'new tradition' of which we have spoken. It exemplified those inter-relationships implied by the word 'townscape' in a way no drawings had been able to do. It brought colour back into the public consciousness, and the play of light, water and the other visual arts as equal partners. But perhaps its real lesson was simply that total design does not just happen, however talented and numerous the team assembled. To bring about such a result, design direction must be professional, trusted and built into the administrative process at the fulcrum point where control can be effectively exercised over the whole field involved, at the right moments of time.

It is a lesson we have yet to learn in everyday life.

6. Product Design

It is of course the boast of industrial design that, harnessed to quantity production, quality need no longer be the prerogative of a wealthy élite but can permeate every corner of our daily lives. Until recently, however, recognition of the designer's role has been more ready in the field of consumer goods than in the field of capital equipment. It is true that the fever of mass selling in an affluent society has led to Queen Anne cigarette lighters and prams that look as though they have been streamlined in a wind tunnel. Nonetheless the range of choice is wide and a latter-day William Morris would find no great problem in furnishing his home with things he knew to be useful and believed to be beautiful. It is not so on the factory floor, in hospital and warehouse, at the docks and on the farm.

The first machines thrown up by the industrial revolution, now the subject of academic study and the most intense efforts at conservation by industrial archaeologists, often had their own sturdy, unsophisticated beauty. The great pumps used in the nineteenth century to circulate London's water supply often exhibit a considerable refinement of form within their, to us, somewhat naïve framework. Early steam locomotives, early motor-cars and early aeroplanes have long endeared themselves to small boys and grown-up romantics. As the demands of sheer technical performance became more complex, however, engineers less and less concerned themselves with visual seemliness. Conversely, whole areas of simpler equipment got left behind and remained almost unchanged for generations. In either case, appearance was either disregarded or made the subject of superficial 'styling'. Although agricultural machinery, for example, is a major British export, lying fifth in the export league of manufacturing industries, with overseas sales at £164 million per annum, it is still dominated by what Professor W. F. Floyd has called 'the blacksmith era of design'.* Relatively little work has so far been done on the ergonomics of farm equipment.

In Europe, it is only in recent years that considerations of efficiency, of prestige, of the need to give a work-force tools and equipment that will allow them a pride in their job, have begun to claim serious attention in heavy engineering and industrial equipment. The United States has for longer taken it for granted. It seems fitting therefore that this story of how a farm tractor was designed should be set in America (herself, incidentally, the leading market for British agricultural machinery). And because it is set in

* 'Men and Machines on the Farm.' *Design* magazine, April 1967.

America, it is also a story of intimate collaboration between the manufacturing firm, Deere and Company, and the consultants they employed; between engineers and designers working alongside one another in a team that shared a single objective. It is a case history, that is to say, which is only in part about a particular range of heavy-duty farm equipment. It is at the same time about the whole process of designing for quantity production generally, from capital goods like locomotives to smaller consumer goods like electric shavers. It is perhaps even more about attitudes of mind.

That the particular project described reflects a consistent and long-standing concern with design on the part of the client company was further demonstrated by the choice of Eero Saarinen as architect for its main administrative centre seven miles out of Moline, Illinois. Set across a dip between two shallow hills and bounded to the south by two lakes; dominated on the outside by its cinnamon-coloured exposed steel structure; enlivened within by constant views of the landscape, furniture designed by Saarinen's office, and two vast murals by Alexander Girard, this, his last, has been acclaimed one of Saarinen's finest buildings and one of the most outstanding of the decade.

Appropriately, the Company's consultant designer is one of the oldest established and most respected in the United States: Henry Dreyfuss. Among the names of those who pioneered the concept and profession of industrial design in the New World, vivid, even extravagant, personalities most of them, that of Henry Dreyfuss has from the beginning stood for sobriety of approach. He has shown a consistent determination to design from the centre outwards, to codify the ergonomic bases of design, and to integrate the design team fully with the engineering and other appropriate departments of his clients. In 1930 he turned down an honorarium from the Bell Telephone Laboratories for styling the 'telephone of the future' on the grounds that no serious consideration was to be given to problems of production and use. (It is nice to be able to note that, as a result of the impracticality of conceptions received from other sources, Bell returned to him and have remained clients ever since.) In 1959 he published *The Measure of Man*—a series of charts of anthropometric data for small, average and large figures, together with a number of equivalent work-place diagrams dealing with reach, sight lines, positioning for maximum convenience and so on. *The Measure of Man* makes no claim to finality but it remains one of the most valuable tools of its kind available to designers.

It was in 1929 that Dreyfuss opened his New York office on Fifth Avenue, after eight years in the theatre, three of them with Norman Bel Geddes. He continued to design straight productions and musicals for Broadway until the industrial design business was established. In the years since then, the jobs tackled by his organization have included such diverse things as magazine st steps, by Henry Dreyfuss and his team, wards a new range of tractors for Deere and mpany. Right, one of numerous quarter-ale sketches, roughly modelled in clay on a oden buck or body that incorporates sential engine and transmission volumes. ncurrently work begins on the seat design; welve-month development programme for s will include mock-ups, like that below, for onomic tests and, later, field trials. Some e years of close collaboration between gineers, designers and production men lie ead, before the first tractor rolls off the line.

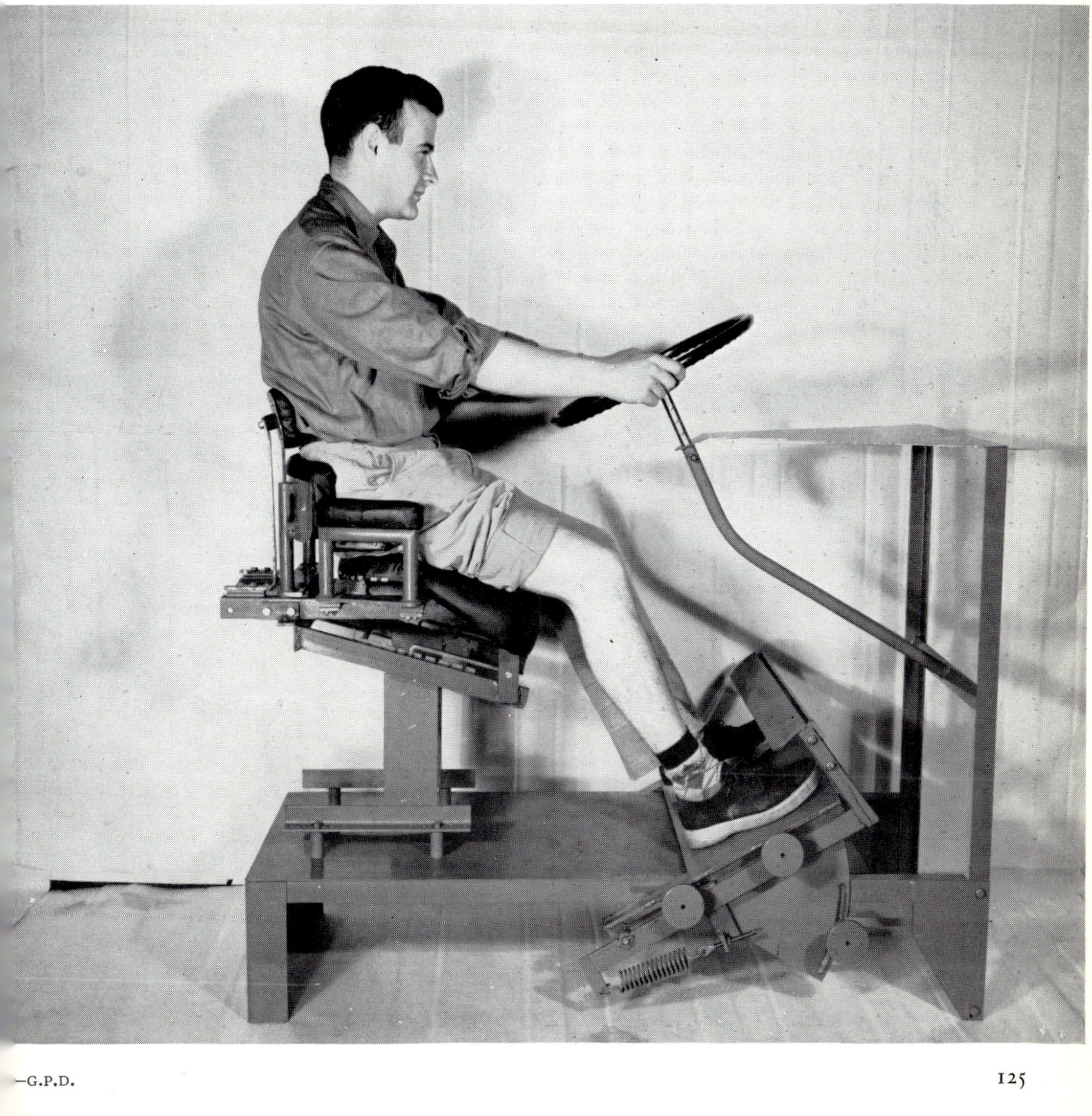

—G.P.D.

JOHN DEERE

l-sized airbrush drawings of the more promising designs, ve left, are made to check proportions. The hood shape of 45 hp tractor, above left, has been greatly simplified since the t sketches but is not yet fully resolved in the 80 hp model. quarter-scale clay model of the latter, left, with dubbed-in re, represents the next formulation and will be used for the t presentation to the client. Once approval in principle is ained for the general approach, further modifications must be de. Full-size mock-ups of doubtful areas are constructed metimes, to save cost, in the form of a mirrored half-model, ight). In the next version of the 80 hp model, above, the front he tractor has been sloped, an additional air intake incor- ated in the front plate, and the visual integration of hood and panels greatly improved. The family resemblance between models is becoming clearer.

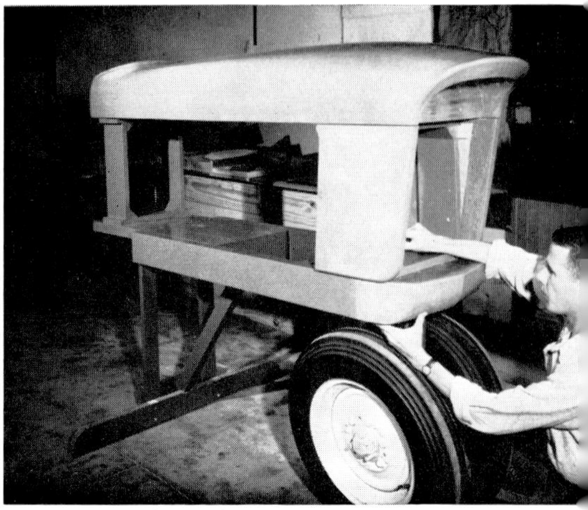

The full-size clay model of the 55 hp tractor hood now takes shape. A wooden buck (top left) is mounted on a reference table on which dimensions are indicated. From full-size drawings, sections are taken at about 5-in. intervals along the centre line, and these are used to cut templates which will ensure accuracy in the clay work. The model is mounted to the height it will be on the tractor. Top right: side panels and frame front casting are now in place. Above left: the model is complete, with air intake grills in place. Field trials have shown that drivers need a sight to steer by, so a small chrome die-casting appears on top of the hood. Finally, above ri fibreglass moulds are taken from the clay, to be ship to the client's factories for use on working prototype The one-piece hood will entail high die costs but wi greatly improve appearance and will mean fewer components to assemble on the production line.

Opposite, top: the control area as finally develope earlier studies for the instrument panel and the whe are seen below. Spokes and hub are minimal, to red masking of the panel, and the instruments themselve have been grouped for maximum visibility.

BE CAREFUL

JOHN DEERE

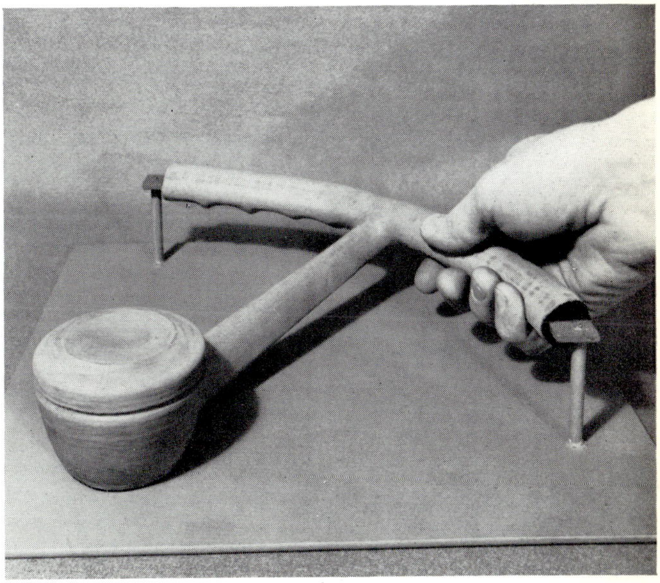

The seat itself, seen here on the 80 hp mode incorporates refinements of a sophistication undreamed of in agricultural machinery only few years ago. It slides on an inclined track to adjust to taller or shorter drivers. When the driver stands or dismounts, a quick-release lever jumps the seat backwards out of his way; when he sits down again a 'memory' device returns the seat to its pre-selected position. The three cushions of the seat are variably sprung and of variable-density foam, according to the part of the body they support; the seat suspension can be adjusted to suit the weight of the driver. Behind the seat, lower picture, every effort was made to clean up the detailing of the various hitches and adjustment points which are inevitably exposed to view.

A full-size mock-up, opposite top, of an almost final design may be compared with the final production model of the 80 hp tractor, opposite below. More years of work lie ahead before the full range will be in production.

JOHN DEERE

JOHN DEERE

Two other items from the same line: a giant 10-ton tractor, above, which powers some of the largest integral farm machinery produced, and below, an example of the related industrial equipment designed concurrently. Heavy-gauge metal, straight lines and break bends in the loader replace the thinner sheet metal and compound forms of the farm units. In all, nineteen tractor models and much ancillary equipmer were brought into production as part of a single, rationalized design operation in which engineers and designers participated on an equal footing.

make-up and exhibition display, plumbing and dentistry fitments; the driver's compartment of a Second World War tank; a rocket-launching platform, tracking gear and control centre; photosetting units; a night club, a Geiger counter and even, it is said, a fly swatter. More currently, the office has worked on steamship, plane, and train interiors, tractors and farm equipment, a mechanical hand for the disabled, telephone sets, filling stations, bank buildings, cigarette-making machines and cameras.

In 1946 he established a second, additional office in Pasadena, California. At the head of the organization is a six-man group led by Henry Dreyfuss himself and William Purcell, his partner, who between them take all responsibility for design. Directly under these are four associates, two in New York, two in California, among whom all the jobs in the office are divided. These associates have the main responsibility for developing designs, meeting schedules and supervising the carry-through of designs into actual production. They fully control the jobs they are handling and may have such help as they require to get them done efficiently. Administration—contracts, accounting, legal matters—has been handled from the outset by Miss Doris Marks.

Total staff, in east and west coast offices together, has never topped 50 and is normally less. 'We feel', says Dreyfuss, 'that this is about all that can be managed in a personally managed profession.' The organization embraces architects, engineers, designers, artists and sculptors, some of whom have been members for twenty-five years. It also boasts a proportionately large clerical staff (it may be noted that Design Research Unit, in London, a comparable office in size and scope, does the same). The organization is regarded as a single unit. A meticulous report system keeps both offices fully acquainted with all details of all jobs (an essential piece of co-ordination, for clients' interests are not necessarily restricted to one seaboard only: for example, the fifty tractors and items of auxiliary equipment marketed by the John Deere Company come from thirteen or fourteen different factories). Nonetheless, commuting between offices is considerable. It has been estimated that staff fly a quarter of a million miles a year. Dreyfuss himself, who 'floats' between the two offices more or less evenly, sometimes spending alternate days in each, clocks up some twenty round trips and 100,000 miles a year.

The number of clients for whom the organization works at any one time is deliberately restricted to about fifteen. These are often, however, themselves very large undertakings. Through the years they have included such firms as American Airlines, the Chrysler Corporation, GEC, Hoover, Linotype, Lockheed, New York Central (railway system), Pan American, Polaroid, RCA, the Singer Company, *Time*, Western Electric—not to mention various government agencies. The wish to avoid increasing the size of the office arises from Dreyfuss's sense of personal service and his wish to develop the relationship with clients in depth, over a long period, rather than to move continuously from one *ad hoc* problem to another. In this he has been conspicuously successful. His first clients, the Bell Telephone Laboratories, as has already been noted, are with him still; several others date back to 1934 and 1935; for Bell the organization has undertaken more than 114 separate projects. 'We are so close to these client organizations' says Dreyfuss, 'that it would be difficult to tell us apart. They visit us, we spend time with them at their factories; we know them well and know what they can produce, what their capabilities are. We accept the mechanical qualifications in their factories as a challenge rather than a limitation. We almost become part of them.'

In the initial research stages of a project, design staff are packed off as a matter of course to the client's organization to study directly the factors which may be involved—behind the counter, in the control tower, in the production department and so on. Interpenetration not infrequently goes a good deal further than this, however. On some jobs there is a frequent exchange of personnel: members of Dreyfuss's office work in the client's engineering department or, frequently, Dreyfuss's organization borrows engineers from the client's staff. Once they borrowed an engineer for an aeroplane job and found, two years later, that they had forgotten to return him.

The same system operates overseas, when the organization is working for American firms with plants in Europe. Precisely the same interchange of personnel takes place—in both directions—as may be appropriate. Deere and Company produced a tractor in Germany of a different horse-power from those produced at home. Dreyfuss ensured that its concept and line fitted those of the rest of the range.

Internally, Dreyfuss retains the right personally to approve or disapprove of designs. In a rare moment of self-justification he says: 'I feel this is one of the things our clients pay us for.' On the other hand, in theory anyway, anyone and everyone in the office has complete design freedom in ideas and on the drawing board. Differences of opinion are satisfied by a meeting of minds: there is said to be very little dictation of design. As part of its internal accounting, the organization has a fund that permits any of its top people to visit Dreyfuss at any time they wish, on either side of the country. This is done fairly frequently, often for a day or two days at a time. Provided he is notified in advance, Dreyfuss will 'clear [his] desk completely and work quietly with that person on a problem until it is solved'.

The thoroughness with which such a firm attacks its design problems is perhaps best shown by following its analysis and development of a relatively simple piece of equipment. As has been seen, the Dreyfuss organization has had a good deal of experience with heavy-duty equipment of one kind and another, from railway rolling stock to battle tanks.

Deere and Company, of Moline, Illinois, are large manufacturers of agricultural equipment. In 1937 they broke new ground when they engaged the services of Henry Dreyfuss as design consultant: the first company in their field to recognize the contribution that industrial design could make to their products. In 1955 the Company took the decision to produce a completely new line of no fewer than nineteen tractor models, together with a host of ancillary items—probably the biggest programme of its kind ever put in hand in this particular field. Henry Dreyfuss and his colleagues were brought into consultation at the outset. The initial briefing and discussion meeting took place in Deere and Company's boardroom in Moline. It was impressively comprehensive, with up to twenty-five people present. On the Company's side were the President; the Vice-Presidents of Marketing, of Product Development and of Tractor Production; the Chief Engineers of the tractor plants; the Project Head, Co-ordinating Project Engineers, the Production Design Engineer and various research men. From the Henry Dreyfuss office came Dreyfuss himself, his partner William F. H. Purcell and a Design Associate with eight years' background in the design of farm machinery. He it was who would have the responsibility, as the project got under way, for all general development and design schedules and for much of the on-the-spot consultation. At Pasadena this team would be augmented by a further designer, one full-time clay modeller with a second part-time, a fibreglass consultant and an ergonomics consultant who would be called in to advise on detailing of the tractor cockpit and seat. All these people had been associated with Dreyfuss for at least five years, some of them very much longer.

The first decisions that had to be taken were general and turned about the degree of innovation that would be appropriate. It was foreseen that development and production would take something of the order of six years and, since a selling life of at least ten years beyond that was envisaged for the range, this meant that the needs and tastes of farmers up to sixteen years ahead had to be judged with some sensitivity. An early move was to undertake research into other existing models. A fairly complete collection of photographs and data relating to American and foreign tractors was made and pinned to 4-foot 'research boards' which were kept handy during all design reviews. Pictures of other then current vehicles—trucks, cars and construction machinery—were also collected to help determine design trends. First preliminary sketches were made to clear the air, help focus thoughts and establish a general direction.

At the same time specific functional points were already being raised. At one of these early meetings Dreyfuss put in a plea to reposition the fuel tank, larger sizes of which protruded, in most then existing tractors, through the main metal hood in an ungainly fashion. It was accordingly agreed that an attempt should be made to mount the fuel tank ahead of the radiator. At the outset the President of the Company had remarked that perhaps the only continuity of design needed on their side was the retention of John Deere's traditional green and yellow paint. With this blank cheque on initiative in one pocket and a general agreement on engineering requirements in the other, the Dreyfuss team turned to their drawing boards.

During the following two months, about a hundred different design approaches were sketched out, and quarter-scale clay models constructed of many. The models were fairly crude and quickly made: rough sketches in three dimensions, with the clay superimposed on a wooden buck in which the basic engine and transmission components were blocked in to give the volume. For the most part, models were only kept long enough to be photographed for record purposes before being cleaned off and remade. Three or four of the more promising were, however, retained for further study and development. Concurrently, full-sized side elevations of these designs were made to enable their general form, silhouette and scale to be studied.

All this time, and indeed throughout the whole programme, there was much discussion within the Dreyfuss group, particularly between Dreyfuss, Purcell and the Associate Designer, who would review the project together and discuss particular problems as they arose from day to day. Written reports of all client meetings and telephone calls were circulated to these three and, since the work itself was done in the draughting office and adjacent modelling shop, there was little difficulty in keeping everyone else aware of the total picture.

The overall conception was now considered sufficiently satisfactory for finished quarter-scale clay models of two sizes of tractor—of 45 h.p. and 80 h.p.—to be prepared for a first presentation to Deere's management and engineers. This, with the preparation of dimensioned drawings and details, and of enlarged photographs of the models with a human figure inserted for scale, absorbed a further month. Thus passed the first half-year. The larger tractor design expressed the vertical fuel tank in front, where a car radiator would normally be, by forming a plain steel plate around the tank. This provided protection as well as a mounting bracket and gave the tractor a massive, purposeful air. The smaller tractor retained the traditional fuel tank mounting position, towards the rear of the engine, to reduce the overall length. The engine was left exposed on the sides for ease of access. The hood was given a curved form which, when seen from the side, would reduce the effect of the hood sloping uphill or downhill when different size tyres were used. The traditional straight top, it was felt, tended to exaggerate this effect, particularly on larger models. Agreement was reached on this general concept and during the next six months a full-size mock-up in wood and clay was constructed by Deere from Dreyfuss's drawings. Members of the Dreyfuss team worked with them for periods of from one to five days at a time, refining the forms and the detailing. Full-sized clay models were made of certain critical design areas. A number of minor engineering changes were made. One year after the project was initiated, the complete full-size mock-up was presented to the management and, with their approval, the first phase of the operation could be said to have been completed. Another four to five years' work still lay ahead.

This work divided itself into a number of distinct phases. Further development had still to be done upon the main prototype tractors; the general design concept then had to be applied to the full range of seventeen models envisaged; a parallel development programme had to be put in hand for the detailing of the driver's seat and cockpit; inevitable minor problems of construction and manufacture had to be overcome when full-scale production was put in hand. During this period the Design Associate visited Deere about every two months to obtain general engineering information and detail. William Purcell took every second or third trip, sometimes alone, sometimes with the Associate; he also would do a considerable amount of consultation at the plant and would be in on major decisions. Henry Dreyfuss himself would be kept up to date by the other two, would be present at major presentations to the client, and would join enough intermediate follow-up trips with his partner or associate to ensure continuity of design quality. Between trips, close contact was maintained with Deere by exchange of drawings and long-distance telephone conversations. This operation, says the Dreyfuss office, went very smoothly.

The second year of development saw some major changes by Deere in the engine design. These required complete revision of the hood form—the most distinctive single element in the tractor's appearance—and new quarter-scale and full-size mock-ups had to be made of this. Below the front of the hood the plane of the tank was now revised to slope forwards and upwards. The severe simplicity of the original design had apparently caused some concern in the Company's marketing group, who had expressed a wish for a rather more 'traditional' appearance. The new line reduced the apparent overhang of the frame beyond the front wheels. It further served, not only to give the machine what was considered to be a more 'dynamic' appearance of pulling, but to provide additional air-intake area as required by the engineers.

While the general design development was being advanced, with members of the Dreyfuss team travelling regularly to the John Deere Waterloo Tractor Works in Iowa, and Deere engineers travelling regularly to California, a second unit was working in parallel on the sub-problem of the driver's seat. This team was drawn mainly from the Company's Product Development Department: an independent group quite distinct from those in the tractor factories already working on the project. To them Dreyfuss introduced Dr Janet Travell, of Cornell Medical College in New York, who had been retained for many years by his office as a consultant on the ergonomics of seating and controls, and had worked with him on many diverse projects.

Until not so very long ago a man operating a farm tractor was perched high upon what was little more than a metal bicycle saddle, which gradually developed into an only slightly less uncomfortable bucket seat. 'Tests showed that bucket seats produced fatigue because only one position could be assumed and undue pressure was exerted under the thighs and on the coccyx.' Rough working over rough terrain demands especially thorough consideration of physiological requirements in order to minimize fatigue.

Dr Travell evolved a basic design concept for the seat unit and the Product Development group then constructed a number of prototypes with numerous adjustments, so that its anthropometric values could be tested and revised. The Dreyfuss office, working with the Deere engineers, then developed its final configuration from the appearance and manufacturing points of view. The final seat design has been described thus:

> The seat adjusts to the physical proportions of the driver. Resting on an inclined track, the seat slides up and back for taller drivers, forward and down for shorter drivers. A quick-release lever jumps the seat backward when the driver stands at the wheel or dismounts. When he sits down again, a built-in 'memory' device returns the seat to the same pre-selected position on the track.
>
> The seat consists of three separate cushions—a seat cushion; one for the lower back (or lumbar region) that fuses with cushioned arm rests; and an upper back cushion. This unique three-part design supports the natural S-curve of the spine. Particularly vital, the lower back cushion minimizes familiar 'low back pain'. The upper back cushion is spring-mounted to give support to the back on smooth terrain when the driver can lean back. On rough ground he sits up straighter and can avoid any pounding effect caused by the high seat back.
>
> The cushioning itself is of 'variable density foam'; firm in some areas, soft in others. This is particularly important in the seat cushion, which is firm in the centre (where the body's weight rests on the bony points of the pelvis) but soft at the forward edge, to minimize pressure on the large blood vessels in the thighs. The seat suspension is by 'rubber in torsion' and has about 4 inches of vertical 'travel' to soak up road shock. A knob permits the driver to adjust the suspension to suit his weight.*

No less thought was given to the control area. The three spokes and hub of the steering wheel were kept as small as possible to reduce visual interference with the instrument panel. Legibility of dials and numbers was given much thought and these were placed to obtain maximum visibility in relation to the wheel. The hydraulic controls are grouped on the left of the steering column, each knob being given a different feel to obviate confusion. The gear shift lever is on the right of the column and the shift pattern is marked by Mylar decals.

* *Automotive Industries,* 15 May 1961. Chilton Publications, Philadelphia. Some other information in this chapter has been gleaned from the same article.

During the next two or three years, a related basic design was developed in the same way for the *industrial* line of tractors. The two designs, one agricultural and one industrial, were then applied to the remaining seventeen tractor models which made up the whole project. All these models were developed in the same general fashion, except that an increasing proportion of the design work—that is to say, the adaptation of the designs to models of different sizes—was done by Deere's engineering staff, who constructed the full-size wooden mock-ups which the Dreyfuss team then reviewed. Critical areas would be remodelled or reworked on the spot. At the same time, detail work was proceeding on the steering wheel and controls, as described, on nameplates, fenders and the design of castings. The back of a tractor is necessarily complicated, but every effort was made to simplify the various hitches and adjustment points involved. More: it is interesting to note that the designers worked on the engine itself. The Dreyfuss team were given a wooden mock-up of the engine and remodelled it in clay to simplify the forms and to ensure that it looked clean when mounted on the tractor (it will be remembered that the engine was to be exposed to view at the sides for easy access). This work was done in the Pasadena office and the Deere engineers came there to discuss the problems involved.

The final design phase, lasting another year, dovetailed into the full-scale production programme. About every two months the designers visited the factory to keep up to date with changes involving the redesign of any visible parts. They also made trips, with Deere engineers, to the tool makers responsible for making the dies used to stamp the large, single-piece metal hood. This, they found, was a critical phase, for the die makers' interpretation of the drawings supplied never looked quite as good as the original models and mock-ups. On-the-spot corrections and refinements were called for and checked again before delivery. The hood—more particularly the forming of its front lip—again caused trouble in the early production phase and Dreyfuss's office proposed an alternative design treatment. However, the problem was overcome by the production group themselves without any basic design changes having to be made. Numerous minor problems of manufacture and production costs arose from time to time, as is normal. If these affected appearance, they would normally be sorted out at one of the regular joint meetings with the designers; if, however, the matter was urgent, a telephone call brought someone from Pasadena to the Middle West by nightfall.

Thus was the whole range of models nurtured into existence. In Germany, a secondary range was produced by the John Deere-Lanz factory at Mannheim, with certain modifications for the German market. The principal differences here were in the curved fenders, which may be used as a seat for a passenger, a legal requirement in Germany, and in the perforated side panels that enclose the engine. The latter was considered to be a national, or

European, market preference: possibly stemming from the use in rural areas of tractors as a general means of transport, with a consequent desire for them to look more 'finished'. Exactly the same interchange of drawings and staff took place between the German factory and the Dreyfuss team as in the case of the domestic American range, and the family resemblance was thus maintained.

The scale of an operation of this kind may be gauged by the fact that no fewer than 110,000 drawings had to be made by Deere engineers over these six years. If the role of the designers in the project can be considered successful, this was, in the words of William Purcell, because 'we were completely integrated with the Company's engineering and marketing groups. The Company brought us into action at the very beginning. Good liaison was maintained throughout between our office in Pasadena and the Waterloo Tractor Works in Iowa. Our role was not cut off as soon as designs were on paper—we were keyed into any changes affecting the final appearance and convenience of the tractors.'

7. The Built Environment: I

In the building, and rebuilding, of our cities is to be found the supreme test of man's ability to work with his fellows to a common end. From ancient times the vision of an ideal city, a 'heavenly Jerusalem', has haunted him. From Sforzinda and Utopia to Owen's 'Villages of Unity and Mutual Co-operation' and le Corbusier's *Ville Radieuse*, the pretty names of regulative models come down to us. Realization, from Palma Nova to Brasilia, has inevitably fallen short of the ideal. At any given moment urban form is the momentary reflection of changing societies, changing needs. There are neither absolutes nor finite solutions in planning. Large-scale architecture in the world today can seldom be truly 'ideal', subject as it is to the most numerous and diverse factors: the shape of the site and the nature of the subsoil; the existence of underground services; the changing uses to which it will be put; plot ratio, daylighting, car parking and many similar, and enforceable, codes; the aesthetic or functional demands of neighbouring buildings; the time available for design and construction, the availability of labour and materials, the cost ceiling imposed. These and scores more limit the range of possible solutions to a mere handful of those potentially possible in ideal circumstances.

It is easier to fault the architectural manners of a new building than to resolve the conflicting factors which urged it into that particular mould. We seek to rationalize the economics-versus-design problem by comprehensively developing whole blocks, whole city centres, at a time: but can we retain in creation on this scale the complex lattice of relationships that emerges from sustained organic growth? A soaring world population is leading us to think in terms of new capital cities, new city regions, new towns and the expansion of old ones. Many nations of the world will have been forced almost totally to remake their physical environment before the end of the present century. By what sort of process will this take place? How does the new shape on the skyline actually arrive there?

This is the story of how two groups of buildings came into existence in capital cities four thousand miles apart. One involved the redevelopment of less than two acres of one of the great cities of the world; the latter was part of a wider project—the creation *ab initio* of a new capital. The former, a group of buildings serving very different—and relatively sophisticated—purposes, was in London. The latter, a group of eight office buildings forming a Government secretariat, was in Islamabad, new Federal Capital of Pakistan. For the former, all those professionally concerned with the planning and design of the buildings—town planners, architects, surveyors, engineers and designers—worked in the same office. In the case of the latter they were widely separated: in Greece, Italy, England, the United States and Pakistan itself, communicating with each other only through the technical code of design. That the single office formed part of the largest local authority in the world cannot disguise the fact that it was a very large-scale group practice of a wholly integrated kind. That the groups and consultants playing their part in the creation of Islamabad were in private practice and widely dispersed cannot disguise the fact that, for this purpose, they constituted a single, multi-disciplinary team.

These two case histories make it abundantly clear, if proof be needed, that there is no magical administrative formula by which integrated collaborative effort may be attained in design. Variations of method are infinite—provided that there exists among the members of the team a common ideology. Lacking this, no conceivable method can enforce harmony. If more space is devoted to the group of buildings in London, that is because it is a type case in miniature of the conflicting demands we have everywhere to resolve in *re*-making our environment, and this is the greater problem. It is of intrinsic interest, moreover, to examine how so vast a body as the London County Council, which contrived to produce buildings of higher than average quality over a period of twenty years, organized so creatively the talent at its disposal. The Ponti-Fornaroli-Rosselli studio with which it is contrasted is tiny by comparison. Here the primary interest lies in how such a practice, itself resolutely multi-disciplinary, formed its working links with colleagues far removed geographically. Both cases, it may be felt, exemplify a commitment to higher standards. Until similar patterns of skill can be deployed more generally, not merely in capital cities but in every town and village, on quite small jobs as well as important and 'prestige' projects, our surroundings will continue to deteriorate.

In 1855 the unprecedented growth of London forced Parliament to take a first step towards rationalizing the ramshackle structure of vestries, parishes, Poor Law Commissioners, Boards of Guardians and the rest, by which London, in common with the rest of England, was still administered. In that year was set up the Metropolitan Board of Works, together with 41 elected district authorities, to provide a number of essential services for London as a whole. Some thirty years later these and other responsibilities were in turn transferred by the Local Government Act of 1888 to what became Britain's youngest county and the world's largest local authority: the London County Council. This creation of an urban county without a gentry, without a tradition, at first met little enthusiasm. From the outset the Council remained severely practical. Over the next forty years, other functions were transferred to the LCC: education, poor law and hospitals among them. The reorganization of the nation's social services and planning legislation after the Second World War yet further augmented the Council's responsibilities. Not all the smaller nation states of Europe could boast as powerful a machine as ticked away behind the 5¼ miles of beeswaxed corridor in County Hall.

The County of London covered 117 square miles and came to embrace a resident population of some 3,250,000—the centre of a larger commuting area of nearly 10 million inhabitants. In 1963 the LCC employed about 73,000 people and its total payroll was said to be the largest in the world. Its annual budget included over £160 million for maintenance expenditure and nearly £56 million for capital expenditure. Within its boundaries there existed the largely autonomous City of London: the old, historic square mile, ruled by the powerful craft Guilds and retained as an endearing administrative anachronism, and 28 metropolitan boroughs, each with its own mayor and council. The division of statutory and permissive responsibilities between the Crown, the central government, the LCC, the Boroughs and various autonomous utilities is not easily summarized. Port, transport, water, gas and electricity were the responsibility of separate public agencies. Certain functions were shared by the LCC and the Boroughs. Foremost among the services for which the County was wholly or mainly responsible were town planning, housing, education, fire protection, the ambulance service, sewage, health, welfare, children's care and the maintenance of certain parks.

The Council itself consisted of 126 elected members, plus 21 aldermen. Business was conducted through 16 main committees, to whom, with reservations, the Council delegated authority to act on its behalf. These were: Children's, Education, Establishment, Finance, Fire Brigade, Health, Housing, New and Expanding Towns, Parks, Public Control, Rivers and Drainage, Roads, Supplies, Town Planning, and Welfare, each consisting of anything from 12 to 50 elected members. In addition there existed the General Purposes Committee, consisting of one member elected by each of the other 15 committees (usually their Chairman) plus 23 members appointed by the full Council. It was General Purposes' particular responsibility to coordinate the views of other committees before their submission to the Council. To this all-important Committee, under the Chairman of the Council himself, would come for ironing out all conflicting claims and policies. Full Council meetings were held fortnightly; Committee meetings normally at similar intervals, or monthly.

The Council's executive Departments followed the pattern set by the Committee framework only partially. For example, town planning and the housing and schools building programme were all administered by the Architect's Department; no Architectural Committee, however, existed as such. The administrative pattern was therefore complex; it may perhaps be best seen as a series of executive units serving a number of 'client' committees and serviced in turn by various technicians and administrators whose advice was at the disposal of the Council as a whole. The total number of Departments was 16. They were those of the Clerk of the Council, the Comptroller, the Solicitor and Parliamentary Officer, the Architect to the Council, the Children's Officer, the Education Officer, the Chief Engineer and County Surveyor, the Director of Housing, the Medical Officer of Health, the Chief Officer of the Fire Brigade; the Chief Officers of Parks, Public Control, School Meals and Catering, and Supplies; the Valuer and the Welfare Officer.

In 1963, exactly three-quarters of a century after the LCC was created, its end was decreed. The *London Government Act, 1963*, transferred, from April 1965, many of the LCC's powers and responsibilities to a reduced number of enlarged and amalgamated Boroughs; the remainder to a new body, the Greater London Council, which covers an area of about 750 square miles and embraces nearly 8 million inhabitants.

One by-product of the Act which caused disquiet was the effect it was bound to have on the integrated team that had been built up within the Architect's Department of the LCC. The Department continues with the Greater London Council, changed in structure but not greatly changed in numbers—and indeed under the same chief Architect—but, shorn of its Town Planning Division and other responsibilities, its role can never be quite the same again. Work of distinction, in housing and schools, had been done by the LCC before the First World War within the social and architectural ideal established by Morris, Lethaby and Webb. Between the wars 80,000 new dwellings were produced but were, architecturally, of an abysmally routine character. It has been suggested that this may have resulted from the hierarchical structure of the Department: that 'the top men were too far from the drawing board'.* The Department began to show its real potential during the Second World War, with the production of the historic Forshaw-Abercrombie *Plan for Greater London*. After the War much damage in the capital had to be made good and, in an atmosphere of intense idealism the Department was built up by several architect-administrators of exceptional talent—notably Robert (now Sir Robert) Matthew, later President of the RIBA, and his Assistant and successor Leslie (now Sir Leslie) Martin. Under them, work was done which received world-wide acclamation and provided a pace-setting lead for other public authorities everywhere. In 1954 the Department, in the person of Sir Leslie Martin, gained an RIBA Bronze Medal for its Ackroydon Estate—the first time such an award was given to a local authority.

* 'LCC Architect's Department', Gontran Goulden, *Architect and Building News*, 31 March 1965.

In its last year of independent life, annual expenditure by the Department on building work amounted to about £24 million. The Department was responsible for the development of London in accord with its own statutory plan, both by control and by such direct measures as were appropriate and possible; for all building work by the Council; for the maintenance of all Council property except housing; for drafting and administering the Building Regulations (including fire precautions); for administering the London Building Act and appointing District Surveyors; for the care of historic buildings and, a nice touch, for street naming. To meet these demands the Department's total authorized strength had grown to nearly 3,500 of which perhaps some 550 were administrative and clerical, and some 2,000 were professional and technical (including site supervision) staff. The latter included architects, planners, structural, mechanical and electrical engineers, surveyors and quantity surveyors, interior designers, modelmakers and photographers.

How was it that out of this vast architectural factory emerged work of such distinction and variety? American practices have reached gargantuan proportions but none has so far developed much beyond the thousand staff mark; beyond a certain point the sheer problem of communication and coordination exerts its own drag. Working for a public authority on the other hand simplifies as well as complicates the job. Committees can, as we all know, reduce design to the lowest common denominator of their members' tastes. Nonetheless, if the will to experiment is there, the closeness of the designers to those administering the by-laws and official building regulations can mean that many legislative snags are ironed out in the early stages of a job; and more smoothly, it may be, than for the outside developer. In the case of the LCC, moreover, the support forces available were exceptional, and far beyond what most architects in private practice can expect. Michael Powell, who was Schools Architect and is now GLC Education Architect, is one of those who readily acknowledge these research and technical resources. Himself in private practice before joining the LCC, he was, he says, much too know-all on his arrival there and wanted to do everything himself. He soon came, however, to realize 'the magnificent facilities available within the LCC to help the architect' and was thankful to be in a position to avail himself of them.

Matthew's gift for choosing men of the right calibre was aided by the temper of the times. Building licences, by which the national post-war building effort was directed into the channels of social priority, were only dropped in the early nineteen-fifties and until then opportunities open to the architect in private practice remained limited. It is nonetheless remarkable that this enormous department, which could have proved unwieldy to the point of stagnation, in fact functioned so fruitfully. Indeed, growing freedom of labour and materials in the mid-fifties served only to improve the quality of its work and it can with truth be stated that the LCC's architectural standards continued to rise almost to the end of its life. That this was so was due in

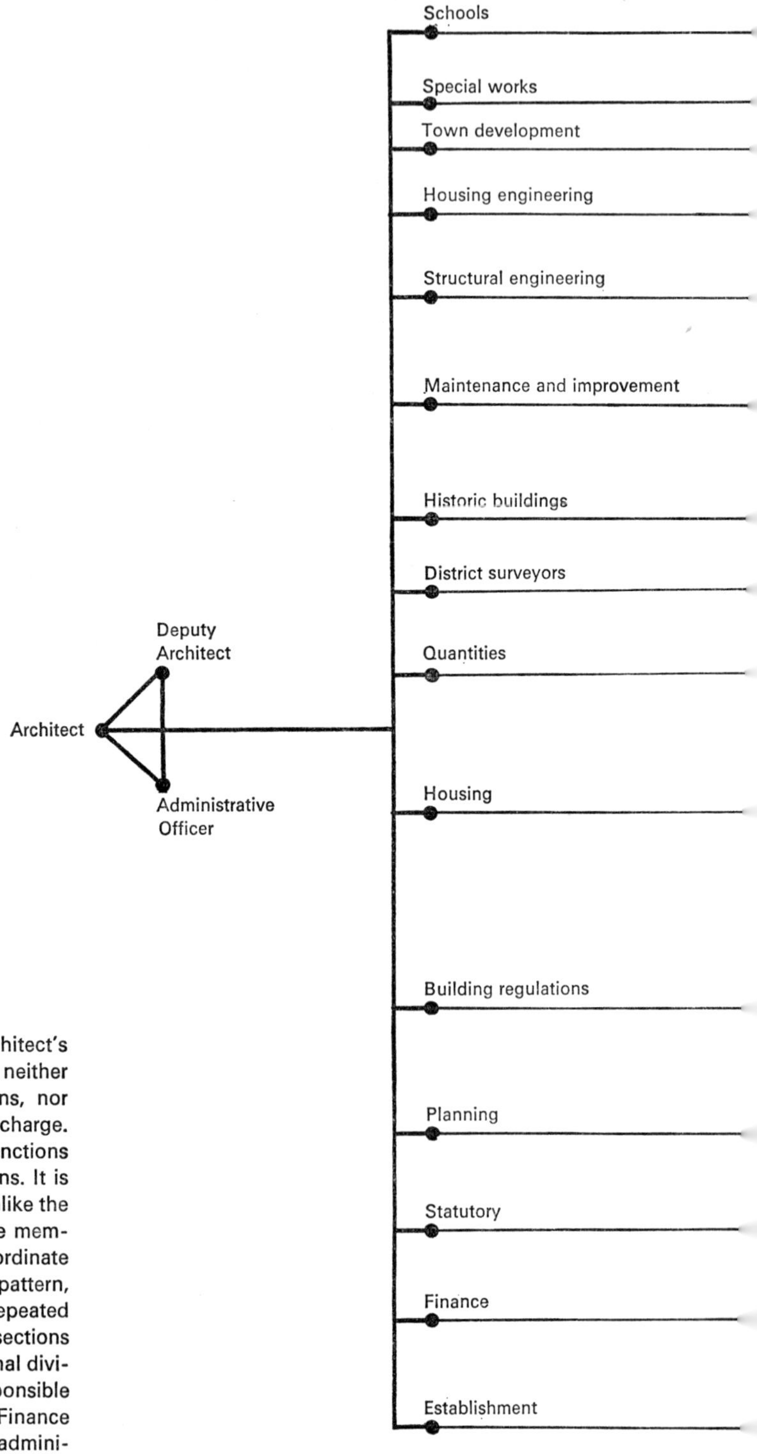

London County Council, Architect's Department. This chart shows neither the numerical sizes of divisions, nor the relative seniority of those in charge. Its purpose is to indicate how functions were allocated between divisions. It is to be noted that in the LCC, unlike the Civil Service, the administrative members of a department were subordinate to its professional chief. This pattern, shown at the senior level, was repeated at all stages. Administrative sections were attached to the professional divisions they served, but were responsible to the Administrative Officer. Finance and Establishment were purely administrative.

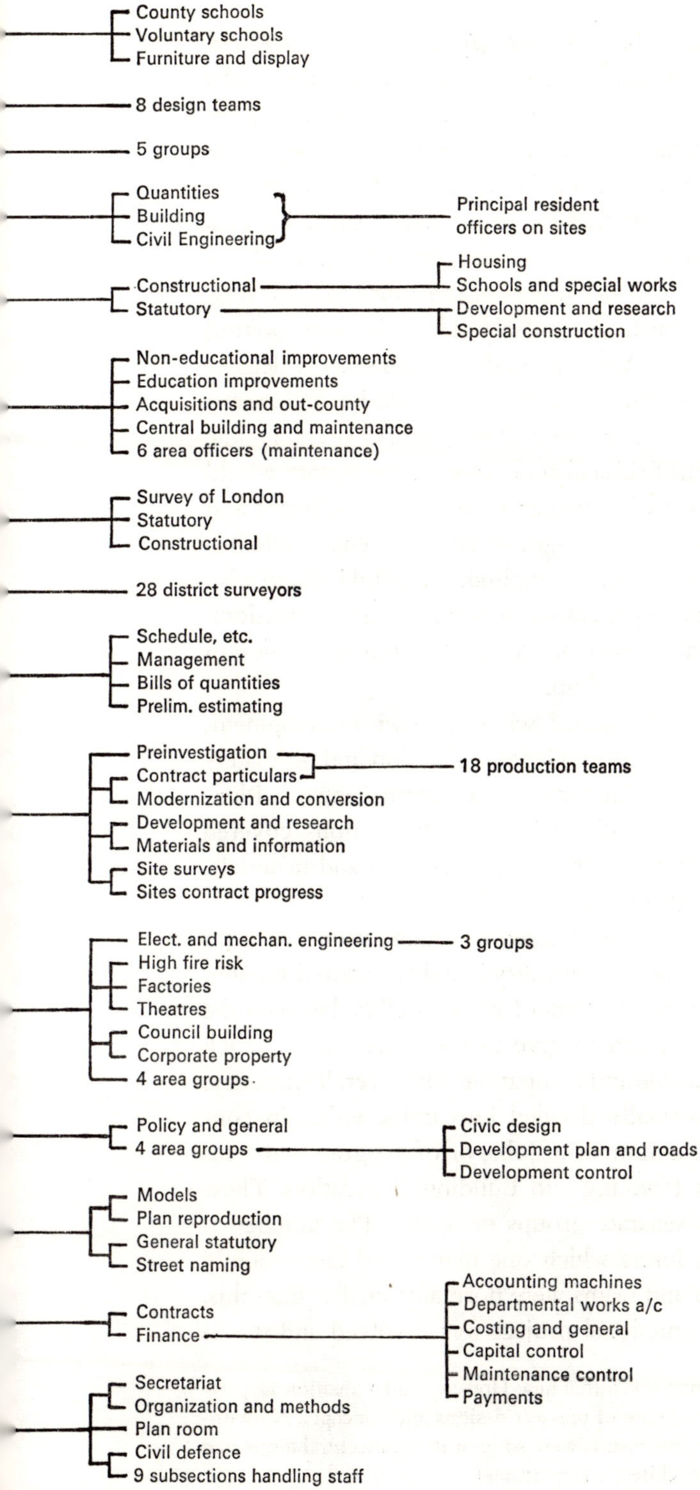
County schools
Voluntary schools
Furniture and display
8 design teams
5 groups
Quantities
Building
Civil Engineering
Principal resident officers on sites
Constructional
Statutory
Housing
Schools and special works
Development and research
Special construction
Non-educational improvements
Education improvements
Acquisitions and out-county
Central building and maintenance
6 area officers (maintenance)
Survey of London
Statutory
Constructional
28 district surveyors
Schedule, etc.
Management
Bills of quantities
Prelim. estimating
Preinvestigation
Contract particulars
18 production teams
Modernization and conversion
Development and research
Materials and information
Site surveys
Sites contract progress
Elect. and mechan. engineering
3 groups
High fire risk
Factories
Theatres
Council building
Corporate property
4 area groups
Policy and general
4 area groups
Civic design
Development plan and roads
Development control
Models
Plan reproduction
General statutory
Street naming
Contracts
Finance
Accounting machines
Departmental works a/c
Costing and general
Capital control
Maintenance control
Payments
Secretariat
Organization and methods
Plan room
Civil defence
9 subsections handling staff

large measure to the manner in which the creative talents within the Department were harnessed.

On taking office, Matthew had abolished the pyramidal structure and substituted what were virtually autonomous teams able to see their jobs through from inception to completion. At the head the Architect, Deputy Architect and Administrative Officer formed a compact directing team responsible for formulating general departmental policy and for communicating it through the network of executive Divisions which carried it into effect. This pattern of a controlling triumvirate was repeated generally in each of the thirteen Divisions with a senior officer of the appropriate profession, a deputy and a senior administrative officer. They, in turn looking upwards, would assist and advise the Architect in the formulation of departmental policy; and then looking down, secure its implementation and execution within their command. There was, throughout, a firm dividing line between the administrative and professional staff (though the former would often defend the work of the latter with greater vehemencc). Elaborate and bulky check forms had to be filled in at all stages of the job to enable administrative and financial control to be properly exercised, but hand in hand with inevitable local authority bureaucracy went exceptional executive freedom. Broad lines of guidance were laid down on design but the actual design process started at the bottom and worked up.

The seven major Divisions were Housing,* Schools, Town Development, Special Works (which split originally from General Division and ultimately more or less reabsorbed its parent), Maintenance and Improvements, Planning and Building Regulation. Specialist Divisions or Sections covered quantities, structural matters, furniture and display, research and materials, historic buildings and half a dozen other fields.

The common principle which governed the whole organization was as far as possible to delegate authority and responsibility in a downward direction, to reduce to a minimum the number of tiers of responsibility between the Architect and the 'coal face' staff, and to give to the individual as much freedom of decision as was practicable and compatible with overall efficiency. Within Divisions, work was normally divided into units, either in constructional divisions, in terms of a number of jobs, or of geographical areas (for example the Boroughs) for Planning and Building Regulation. These units of work were handled by separate groups or teams. The number of staff in a team was kept within limits which one man could conveniently control; twenty was a maximum and teams were normally smaller than this. Where a particularly large constructional project was involved and it was

* Immediately after the war housing was given to a new Housing and Valuation Department which, to speed construction, made use of pre-war designs and concepts. Notwithstanding dramatic numerical successes, the results were so poor in architectural terms that in 1950 housing was returned to the Architect's Department.

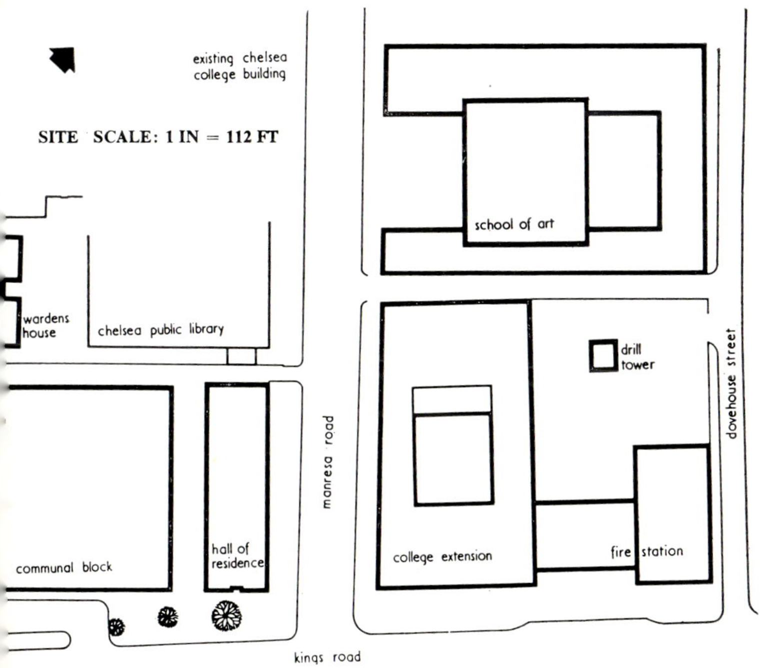

Redevelopment in King's Road, Chelsea: a fire station; a science block for some 330 students and staff for a College of Advanced Technology (now part of London University); a hall of residence for 186 and, just visible at the left, a block containing assembly hall, common rooms and a canteen for 800. Out of sight lies yet another building type: a school of art with 220 students and some 40 staff. This complex, by the Schools Division of the LCC Architect's Department, presents a unity which is remarkable in view of its diversity of purpose. The site, split by Manresa Road, covers only 1·8 acres.

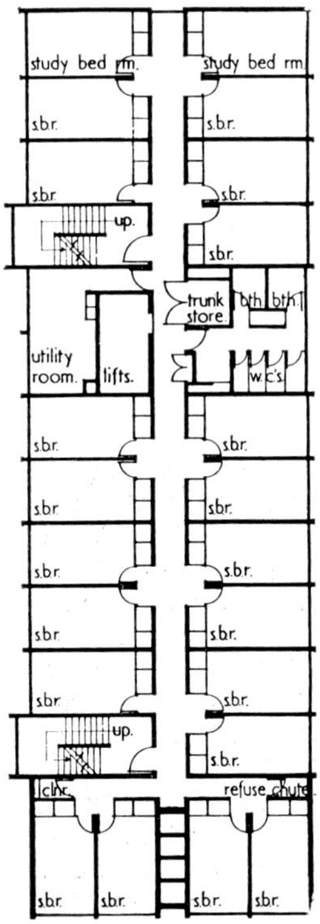

HALL OF RESIDENCE
FIRST FLOOR SCALE: 1 IN = 32 FT

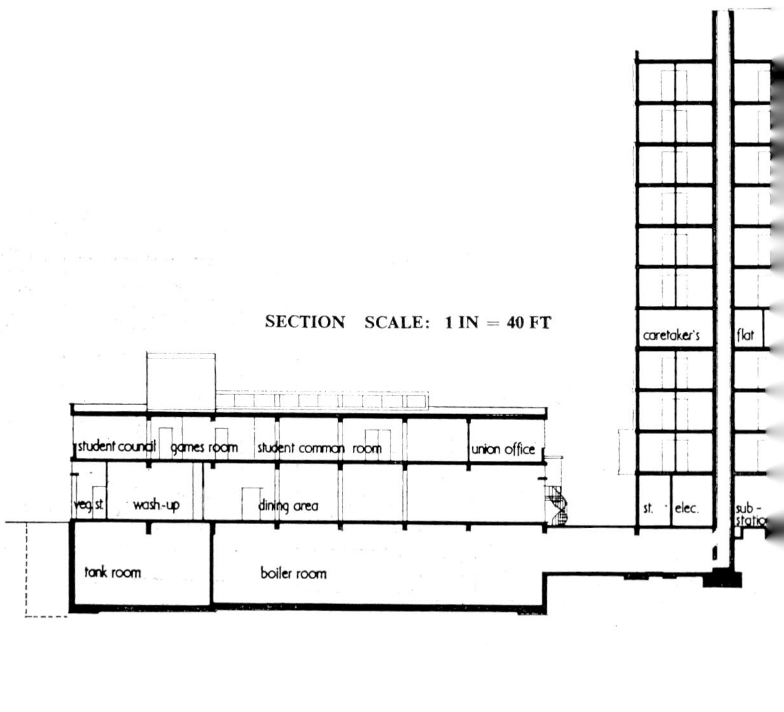

SECTION SCALE: 1 IN = 40 FT

is was how the whole thing began, facing page, top: h sketches of the area by some of the architects cerned, to help them get the feel of its character. st straightforward of the blocks was the hall of idence, above, the requirements of which more or s forced it into its final form and size. Opposite ow, plan of the first floor (one of the study-bedrooms is seen at the foot of this page) and a section through it and the adjoining communal block. Above, the courtyard between two buildings (the spiral staircase provides a fire escape from a projection box) and a view from Manresa Road which may be compared with the sketch of six years earlier. Below right, the caretaker's cottage and garden.

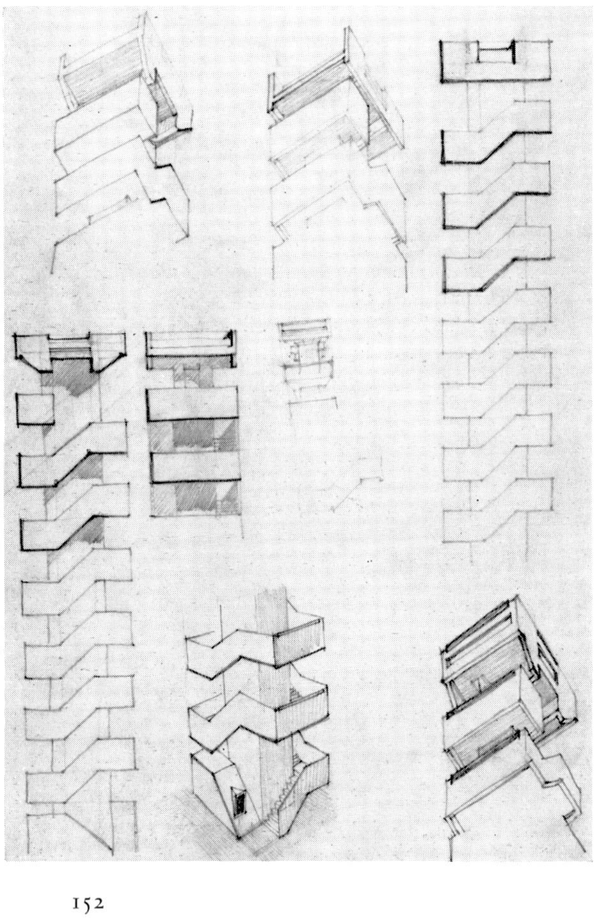

It soon became clear that one of the elements mo likely to clash with the hall of residence was the d tower of the fire station, since this, conditioned b fire service requirements, would be unrelated in height or siting. The model, above, shows the buildings involved; the hall of residence is at the extreme left, and the art school at the right. The tower was developed into a feature in its own righ a kind of square spiral, as the sketches on the lef show, treated boldly and centred on Oakley Stree

bove, the completed tower seen from Oakley Street; e upper part of the art school appears behind it. low, looking south-west across Dovehouse Street from the churchyard of St Luke's. Tucked into the flank of the fire station is an office for the District Surveyor.

The school of art was to be the first in Britain for many years to be designed primarily for the teaching of painting a sculpture. Much research was done, by the job architect, R. W. Robson-Smith, to establish desiderata for the lighting and internal layout of studios. Existing examples were photographed-traditional, as at Camberwell Art Scho top left, and more personal, as is Henr Moore's top-lit studio, centre left. Sim models were made and objects photographed within them, below left, to tes the quality of light that might be expec from an overhead system without othe windows. Advantages were seen in th flexibility and control which such a system offered.

An early sketch, above, may be compared with one of the painting studios as built, below right. Below left, the sculpture studio.

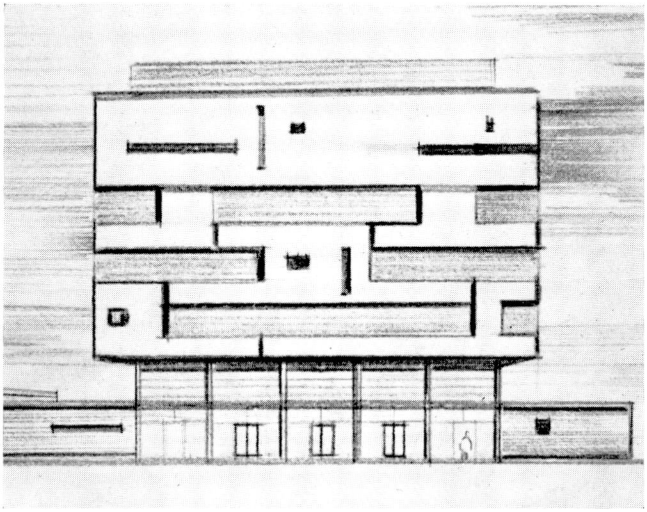

Stages in the development of an idea. The general disposition of the interior spaces of the art school has been evolved. Top and bottom floors will be of double-storey height, the former, accommodating the painting studios, almost entirely without windows; the latter, almost entirely glazed to reduce the apparent volume of the building. First sketch elevations are bold and uncompromising. After all, it *is* a school of *art*.

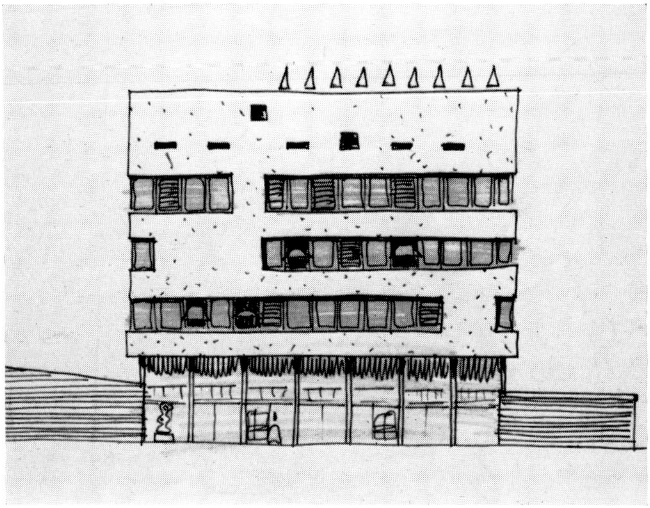

Maybe, though, that *could* look a shade brash for Manresa Road? Perhaps something more disciplined, without actually getting too classical? Work on the lighting of the studio is bearing fruit and the characteristic saw-tooth skyline appears. Not bad. But it is no good pretending it bears much visual relationship to the rest of the development.

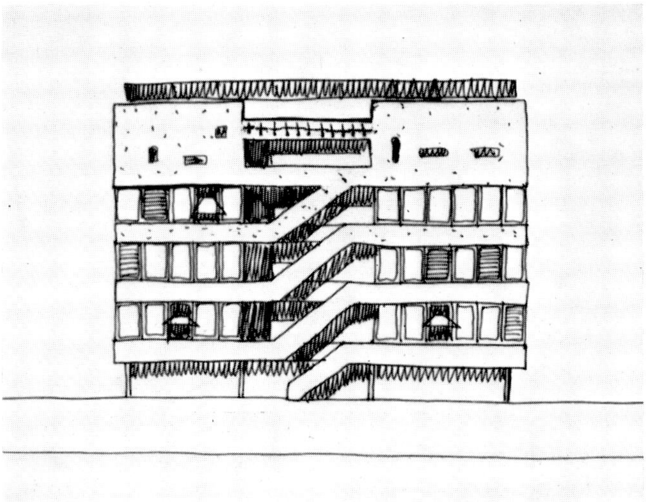

Wait a minute. If the secondary staircase round at the side, in the south-east elevation were expressed vigorously, an accord could be reached with the chunky diagonals of the fire station tower. Or would such a strong sense of movement be too much of a good thing altogether in what anyway has to be a fairly tight-knit group?

ry knocking the staircase back; suppress it a t. This is getting nearer to the scale of the all of residence, but it does seem to lack stinction. What would happen to the front evation?

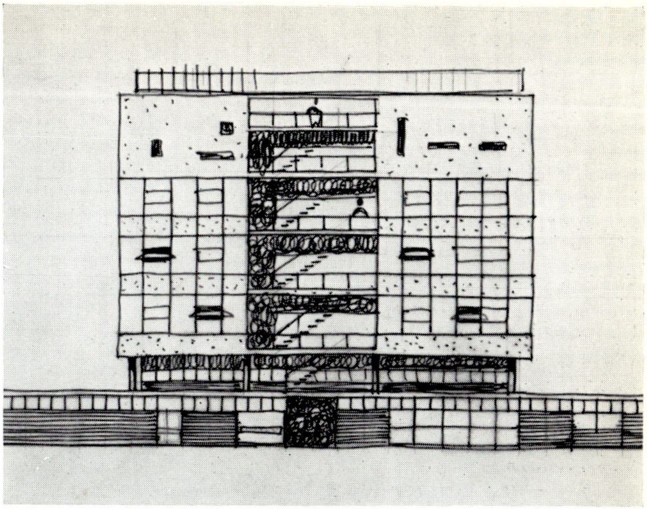

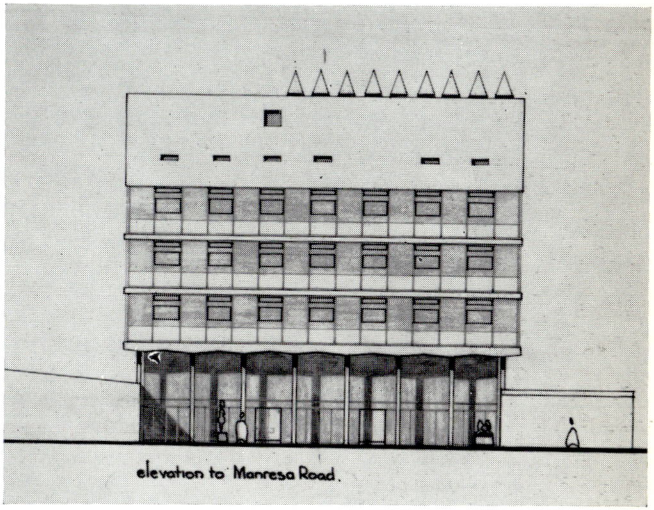

ep the top floor very calm and deliberate. ick to the same sort of infill panels as ross the road. It certainly looks less enuous. And with a little accent to the line er the entrance it would not look too boring.

does need beefing up though, it is just too ined altogether. Go back to square one. ng back the horizontal emphasis. A really ong differentiation between solid and void uld still tie it in with the fire station and the ence block. Maybe express the main paint- studio in the façade as well as by the roof hts. Pull in the glazing behind the pilotis way we had it to start with, to make the in volume float more lightly. That begins to k like something . . .

n fact (see overleaf) the final design was n calmer and more deliberate. The two top reys were broken only by two small tilation slits; the roof lights run symmetric- across the whole elevation. The 'feel' of building had been established, however, final modifications were worked out hin this context.

Opposite page, Chelsea School of Art from Manresa Road. It is possible to glimpse, through the glazed entrance hall, the courtyard at the rear with its pool and its big reclining figure by Henry Moore, right. Below right, the ground-level common-room in one of the single-storey wings flanking the forecourt. Finishes throughout are simple, at times even spartan, but that this in no way inhibited design may be judged by the main stairway, opposite foot, which rises from ground level through the centre of the whole building. Below, a section; the underground car park does duty for the whole redevelopment.

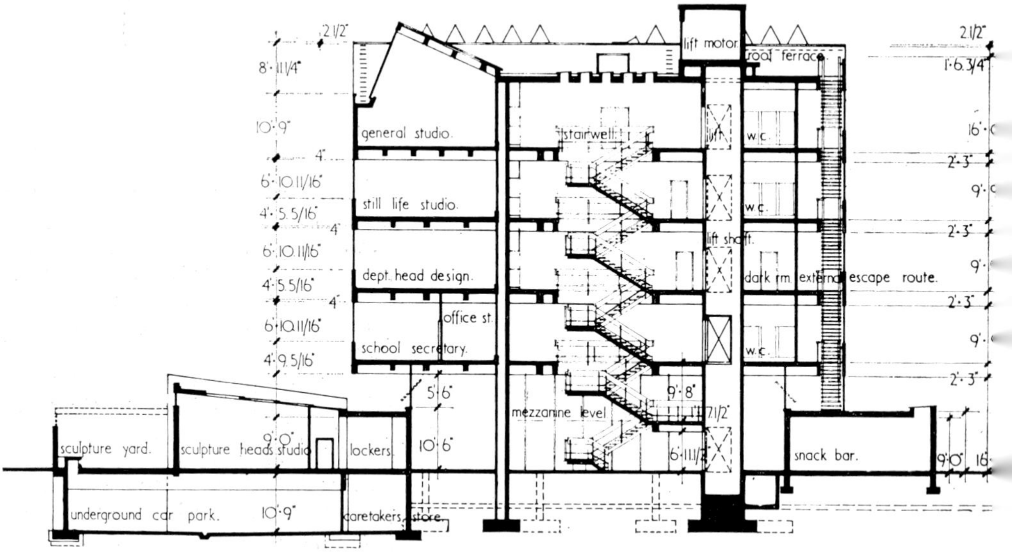

ART SCHOOL SECTION SCALE: 1 IN = 32 FT

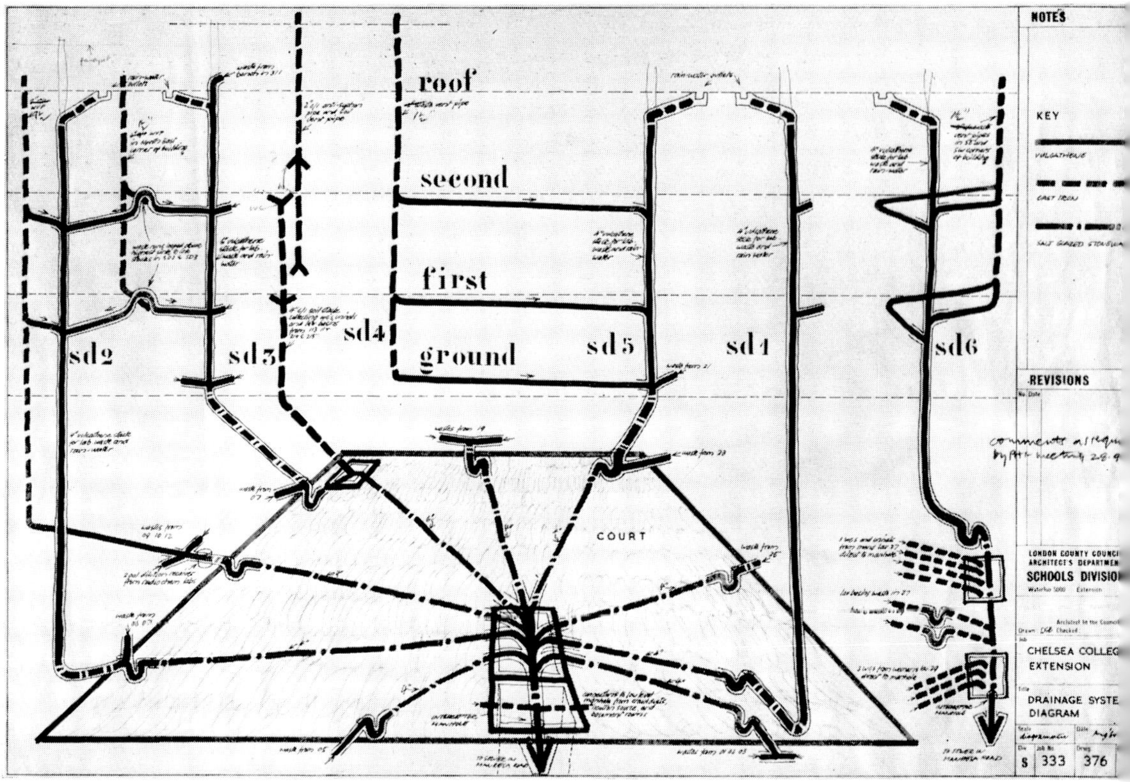

To the passer-by, simplest in appearance of all the buildings in the group is the science block. To the designers it posed complicated technical problems of interior planning. Shown here is a drainage diagram, one aspect only of the complex services system demanded by such a building. As the requirements of science and technology proliferate and change at ever greater speed, so does the need for flexibility in laboratory planning. Functionally, the ideal is to establish a modular space-frame within which ducts and services can be shifted and adapted as required. In Chelsea, however, the issue was further complicated by budgetary limitations and the code of elevational good manners imposed in the interest of the group as a whole.

practicable to divide it, two teams might work jointly on the whole project, although this brought problems in the co-ordination of detailing. Individuals within teams could be transferred freely according to the allocation and pressure of work, so that there was considerable fluidity in the numbers within teams; in broad terms efforts were made to programme work loads so that movement between teams could be kept to a minimum and the impetus of competition between groups maintained.

With programmes of this magnitude it had been possible for many years to support research and development and other specialist groups. Such groups not only helped to solve problems directly associated with the Council's own programme but were able on occasion to add to the general store of knowledge in the field concerned. Basic research and materials investigation were tied to the Housing Division, since this provided the main opportunity for such activity, but information was disseminated and common problems discussed at monthly Development and Materials Committee meetings, under the Architect or Deputy Architect. Each Division was represented on this Committee, as well as the Scientific Advisor whose laboratory and practical testing techniques and assistance formed an invaluable part of the research programme.

In 1950 the Department had two general purpose designers; in the last years of its life the number had grown to 33. This Furniture and Display Section worked directly to any of the client departments who chose to make use of it. The greater part of its work, however, was done for the Education Officer and the Section was therefore placed with good reason within the Schools Division. The importance of its role may be judged by the fact that several million pounds' worth of furniture was manufactured for the LCC every year. Since a good many LCC designs for school furniture were used by other Local Education Authorities in Britain as well, the anthropometric research done by the Section helped to influence the whole field of school furniture design.

The Section, under a Senior Designer, was divided into three working groups: the *Furniture Design Group,* to which were attached special advisory panels with the power to approve or disapprove new designs; *Interior Design Group*—itself subdivided into a *Technical* section which worked on large contract schemes and a *Colour and Furnishing* section which was used by all LCC Departments on redecoration schemes; and the *Exhibition and Graphic Design Group,* used particularly by job architects to co-ordinate signs and lettering in new schemes and to handle the constant flow of exhibitions illustrating the work of the Architect's Department, and indeed, the whole Council. The Section thus exemplified within itself the general group structure of the whole Architect's Department. At the same time, it remained a fairly close-knit unit which might well have to deploy all its resources at one time or another on a single job.

Thus, to take a hypothetical case, the Exhibition Group might find itself called upon to prepare a small exhibition for display at a Borough Town Hall, to explain to local residents the nature of a complex redevelopment scheme proposed for that area. Within this scheme was to be included, let us say, a further education college to be built by the LCC itself. For this the Technical section of the Interior Design Group would probably be called into consultation at an early stage of discussion and would thereafter be kept in touch by regular progress meetings with the job architect and any appropriate specialist consultants. The Technical Section would feed back to the job architect the results of their own research into ascertained user requirements, in time for these to be included in his preliminary planning. As the job progressed, detailed design studies of necessary built-in equipment would be undertaken for every room and a guide price would be obtained for inclusion by the quantity surveyor in his overall estimate for the college. Later again, the Furniture Design Group would probably be called in to advise the Education Officer on all the free-standing furniture and equipment required. Finally, it might be (the choice would be the responsibility of the new Principal) that Colour and Furnishing would be asked to undertake, with the job architect, the decoration schemes for all or some of the building.

What has here been described at some length must be seen as only part of a process which included equivalent contributions by planners, engineers, surveyors, landscape architects and all the other skills involved in environmental design. The pyramid of smallish units of about a dozen men, building up through the sections to the main Divisions and the Architect to the Council himself, thus took unexceptional and traditional military form. What transformed it was the degree of autonomy granted those smallish units. In a machine of this size, generalizations are dangerous. Power pyramids may arise within a group system by reason of the actual personalities involved; amongst so many, not all can be equally forceful, equally articulate, equally gifted. Nonetheless, the LCC Architect's Department was emphatically *not* authoritarian in the le Brun sense. Matthew and Martin are themselves designers, not merely administrators, and their concern with design standards was transmitted to those under them. It was a concern with quality, however, rather than with any preconceived design approach and it could, indeed, be argued that enthusiasm within the groups was not always subordinated to the greater unity as it should have been. Buildings as admirable as the Festival Hall (in any case now extended and altered) show evidence in their detailing of too many, not wholly co-ordinated, minds at work.

On the other hand, the greater part of the Department's work in these years matched execution to conception, soundly in general and sometimes superlatively. Co-ordination of the team obviously existed. Formally this was effected through a system of regular conferences at all levels. The Architect or Deputy Architect and Administrative Officer met their opposite numbers at Divisional level to discuss progress and policy at, generally, monthly intervals. The Divisional heads in turn met their group counterparts, separately and together, at varying intervals; apart, of course, from *ad hoc* discussions on specific problems as they arose. Within the group, methods of consultation varied according to the circumstances and personalities involved. Further cohesion was lent the Department as a whole by the fact that administrative staff at all levels worked through their own chain of command to the Department's Administrative Officer.

At lower levels, discussion would normally turn only about design problems, programming and progress, for it must be remembered that other appropriate Departments would have taken part in the shaping of the brief which therefore came to the designer as an agreed and clear-cut document.

Thus, for example, a brief received by the Housing Architect would be based on a site which he and the Planning Officer had already considered for suitability for residential development at a stated density. The brief would specify the Director of Housing's requirements as to the proportion of dwellings of different sizes. The number of shops would be given, together with other vital information such as road widenings, or new main road proposals, to be safeguarded; areas within the site to be reserved for other services (e.g. welfare homes, children's homes); and ancillary housing and other buildings to be provided (e.g. local management offices, school treatment clinics, maternity and child welfare clinics, day nurseries and so on). The task set the designer might be complex, but he could proceed with a clear knowledge that the requirements he would have to meet in the scheme had already been agreed with, or would be prepared in close liaison with, his town planning and other service and consultant colleagues. An example of interest because of its unusually varied requirements is the group of buildings erected by the LCC at the junction of the King's Road and Manresa Road in Chelsea, and completed in the summer of 1965 at a cost of over £1 million.

The road is narrow and busy with traffic. The trees of the old parish churchyard of St Luke lean outwards towards the light. You are hardly aware of the grey and white buildings on the northern side until you are upon them. First to catch the eye, inevitably, is the eleven-storey slab, at right angles to the main road and much the highest building in sight. Flats perhaps with offices and show-rooms adjoining? But if you should be coming up Oakley Street from the south, you cannot miss the chunky vertical structure that is centred upon your line of approach. That has a less familiar look. From little Dovehouse Street, a low roof with a scarlet cube on top, like a Christmas present gift-wrapped, leads the eye to a four-square building with sawtooth skyline topping a deep band of windowless wall. Across an enclosed courtyard, behind the chunky tower, lies a building with even fewer windows, secretive even from the front, for the ground floor presents only a clerestory light to the King's Road. From Manresa Road if you look across a courtyard, through the building with the zigzag roofline, you may catch a glimpse in a further courtyard beyond of a massive sculpture by Henry Moore.

The King's Road is about as informal as you can get, architecturally. Off it there are handsome squares and terraces, but the Road itself seethes with the gay and with-it bustle of boutiques and tiny restaurants, antique shops and decorators. No call here for Beaux Arts pomposities. On the other hand this diversity of structures on a divided site, which has so clearly been developed as a unified scheme, cannot but intrigue. What links the group?

The buildings do in fact comprise an odd rag-bag of functions: various extensions for a new College of Science and Advanced Technology, including chemical laboratories of some complexity and a Hall of Residence with 186 study-bedrooms; a fire station for four appliances and a strength of 75; a new school of art; and several subsidiary offices and houses. The Henry Moore sculpture, the metal lettering over the doors of the fire station, the purpose made furniture in the chemistry block: these represent only some of the final touches to a total design concept, the first elements of which had emerged some fourteen years earlier and which had been shaped, for better or for worse, to its present form by innumerable interlocking decisions, taken by several hundreds of people, individually, in groups and in committees, during the intervening period.

In this process three main stages may be distinguished. First, the means by which these particular functions became identified with this particular site; second, the process by which the various schedules of accommodation, the requirements of the site and the permissible cost of the project were resolved into a first sketch scheme; and third, the detailed working out and construction of the project in the light of such amendments and modifications as were then seen to be necessary.

Early in 1951, as a result of requirements tabled by its Education and Fire Brigade Committees, sites were approved by the London County Council for inclusion in the Draft Development Plan for an extension to the Chelsea Polytechnic and for a new fire station in the same area. Other sites had been considered, for everyone knew that those proposed were too small and everyone, except the London Fire Brigade itself, would have liked to have got the fire station away elsewhere. The facts remained that the Chelsea 'Poly' was where it was and that, from an operational point of view, this was the right place for a Chelsea fire station. Such a matter, involving conflicting claims by different LCC Committees, would come before the General Purposes Committee under the Chairman of the Council, there to be decided, in effect, by him personally. However, the most pressing need in educational building at that time being for primary schools to meet the first waves of the post-war population 'bulge', it was not in fact until 1955 that the Ministry permitted the Polytechnic extension to be included in a draft building programme. The Polytechnic had an architect of its own choosing in mind but the LCC's Finance Committee felt that greater control could be exercised if the work were done by the Architect's Department, and thus it was agreed.

By now, however, wider implications had become apparent. The Government was becoming aware, on the one hand, of the need nationally to expand the provision of places for science students, and of space for advanced laboratories; on the other hand, of the need to rationalize the nation's network of art schools by pooling their resources on a regional basis. The majority of provincial art schools are still housed within the polytechnic or technical institutes built at the turn of the century. Most such institutes having already developed their sites to capacity, the neatest solution to a double problem seemed often to be to build a separate school of art, which can more easily function as an independent unit, and to utilize the space thereby freed within the institute for the expansion of scientific facilities.

A survey on these lines was carried out in London by the Ministry of Education in conjunction with the Education Officer of the LCC. As a result it was proposed that London's public art schools should be reduced to six and, amongst other steps, the Ministry suggested the amalgamation of the schools attached to the Chelsea and the Regent Street Polytechnics in a new building in Chelsea. At the same time, following a Government White Paper on Technical Education, the previous proposals for the extension of the Chelsea Polytechnic were revised to enable it to meet the Minister's requirements for the newly defined category of College of Advanced Technology, one of which was the inclusion of a hall of residence. The revised proposals inevitably required an extension of the site as originally defined and, in turn, an amendment to the County Development Plan. Before this could be prepared, however, an application was received from the Cadogan Settled Estates Company, landowners of considerable areas in Chelsea, for permission to erect luxury houses and flats on part of the proposed education site now required.

During the negotiations which followed between Cadogan Estates and the LCC, the Company expressed misgivings about the proximity to their own development of so many students in residence, quite apart from the loss of land entailed. As a result of certain adjustments, agreement was reached in November 1957 on the sites to be acquired by the Council for the (now designated) CAT, the new School of Art and the fire station. The site was still too small, and the CAT never got as much space as it wanted.

As finally established, the total site lay on the north side of King's Road in two irregular areas bounded by Dovehouse Street to the east and the not inelegant backs of the houses in Carlyle Square to the west, being split down the centre by Manresa Road in which still stands the original Polytechnic building.* On this site, cleared of its mixed collection of worn-out terraces, back-yard workshops and studios, would have to be reconciled the demands of three main briefs from different 'clients', together with a new but subsidiary demand from the Establishment Committee of the LCC for a suite of offices for the District Surveyor.

Collectively, these briefs began to look intimidatingly diverse. The college extension would require living accommodation for about 200 students, dining and common-room facilities for the entire college, a new building for the chemistry department and a house for the warden. The London Fire Brigade have many detailed requirements affecting the design of fire stations, mostly aimed at making it possible for an appliance to leave the station, fully manned, within seconds of the 'bells down' alarm. Those most closely affecting the external layout of the whole complex were the need for the usual drill yard and tower for hose drying and ladder practice, and the categorical stipulation that the appliance room must face Oakley Street, on the opposite side of King's Road, for a quick get-away to Chelsea Bridge and the Embankment route along the Thames.

The new school of art was to be a self-contained unit with its own assembly and dining hall, kitchen, library and administrative offices. Schedules of accommodation for art schools are first drawn up tentatively in accordance with the Ministry of Education's *Building Bulletin No. 5* and are then submitted to the Ministry for consideration. Since this would be the first of its size in Britain since the war to be specifically designed for teaching fine art, it was early foreseen, however, that considerable research would be called for to establish desiderata on finer points of layout and lighting.

Finally, there was the little matter of the District Surveyor's offices while, of course, somewhere on the site car parking—and this was to prove a bone of contention—would be required for the whole complex. The total acreage for all this was only 1·8.

This, then, was the nature of the problem presented to the Architect's Department of the LCC in the autumn of 1957. Inevitably, it came to be handled by the Schools Division. Michael Powell, Schools Architect, himself made some first sketches just to see approximately what accommodation could be got on the site and to make consultation with his 'clients' easier. With the same object, a very rough block layout was evolved. The site, it was clear, would of necessity have to be developed to its full plot ratio; there was a height restriction by the Planning Department of 100 ft.; immediately to the north, Cadogan Estates were putting their own very 'traditional' housing

* At the time of its design the entire scheme thus came under the general aegis of the LCC. When the Chelsea Polytechnic achieved the status of a CAT, the LCC ceased to be the responsible Education Authority, but this change only took place after the contract had been let and affected the buildings only in detail.

development in train. 'The job', says Michael Powell, 'had all the ingredients of an architectural dog's dinner and my immediate worry was to see to it that this did not happen.'

In Schools Division there were three Sections: County Schools, Voluntary Schools and, as we have seen, Furniture and Display. The former at the beginning of 1963 had £11¼ million of work in contract and about £7½ million in various stages of preparation. The whole Department numbered upwards of 70, split into seven design teams, or groups, each of up to 15 'units of staff' (in the quaint mandarin of administration) including up to ten architects, under a group leader. Powell deliberately sought to keep groups to this size ('you can't go above this and still know what's going on'), and to feed them with a variety of jobs so that they did not find themselves specializing. The Chelsea complex, with a building contract likely to be around the £1 million mark, meant either expanding a group to above Powell's optimum size, or sharing the work between two groups. He decided on the latter course. He made one group responsible for work to the west of Manresa Road; another for work to the east. This was a calculated risk, in that a clash of architectural thinking between the groups was an obvious possibility, even though the teams and group leaders worked in the same room so that consultation could be easy and frequent. As the job progressed, however, the groups were themselves to be subdivided, so that eventually the four main blocks were all handled by separate teams, linked by the two group leaders to Powell and, through him, to the rest of the Architect's Department.

By December 1957 schedules of accommodation for the hall of residence and the art school were approved. By March 1958 a sketch scheme for the new buildings was ready for presentation to the three LCC Committees concerned, after informal discussions with the Chelsea Borough Council on traffic and car parking matters.

The layout of this sketch scheme had been conditioned by two main factors: the Fire Brigade's insistence on their site opposite to Oakley Street; and the desire expressed by Cadogan Estates that the hostel should not abut their new housing on the northern boundary of the site. Juggling the areas required for the different buildings around these two fixed points, the layout came out with the fire station in place opposite Oakley Street, with its drill yard and drill tower behind; the new chemistry block alongside on the corner of King's Road and Manresa Road, and the school of art to the north running right across between Manresa Road and Dovehouse Street. The hall of residence faced partly on to King's Road, but mainly on to the west side of Manresa Road opposite the art school—well separated from the Cadogan Estate. The communal block adjoined it, running north from the King's Road frontage. This basic layout was to survive into the completed scheme.

The hostel had to be designed to the maximum permissible height of 100 ft., in order to achieve the required number of study bedrooms and space for domestic staff. It was planned at right angles to the King's Road, so that the majority of the bedrooms would face east or west, and in some cases south (none towards the Cadogan Estate). The height of the building helped to keep the ground floor open and thus to avoid a frowning cliff façade in Manresa Road. The assembly hall was away from the main road, the rest of the communal block—dining-room, common-rooms and the rest—on the main frontage.

Other elements fell into place without undue difficulty. Car parking would be provided under the art school. The District Surveyor's offices would be slotted into the rear of the fire station. The warden's house in the rather secret north-west corner of the site could even be given a little garden.

An awkwardness immediately became apparent, however, in the height relationship of the main elements. The drill tower, to meet technical and training requirements, *had* to be 80 ft. high and sited in the open area behind the station so that vehicles could manœuvre around it. With it had to be accommodated the 100-ft. hostel, a two-storey communal block and the art school—which was seen as the equivalent of seven storeys in height. The drill tower was sited on the axis of Oakley Street, to provide a terminal feature and clearly to centre the fire station visually on its main access route. At this stage, however, no one felt very happy about the three-dimensional composition that was forming. Nor was it easy to see how the various 'envelopes' could be greatly modified. The official plot ratio* had already been exceeded and, although it seemed that this might be permitted for the educational purposes here involved, the designers were warned that if they allowed the hostel to rise above 100 ft., residents in a surrounding area of such and such a size could legally appeal: and in that part of Chelsea would probably win. This sketch layout was nonetheless agreed in principle by the various 'clients' concerned. It then went to the LCC Town Planning Committee for approval and from them to the Royal Fine Art Commission, who also approved it in principle. Now began a long period of formal consultations, costings, research and modification as the buildings were designed in detail.

It was at this stage that the architectural team was split into four groups, each responsible to Michael Powell through the leader concerned. One was made responsible for the communal block and warden's house; one for the hall of residence; one for the school of art; one for the fire station, chemical block and District Surveyor's offices.

The Principal of the Polytechnic and his Heads of Departments were consulted on all aspects of the CAT buildings: layouts and fittings to laboratories and the administrative problems of laundry and refuse disposal. When these matters had been finally thrashed out, the results were submitted to the Governors for their approval. The same process was followed with the fire

* Ratio of total floor area to the area of the site.

station and the school of art. In the latter case, however, certain additional considerations came into play.

When the proposal to amalgamate the two schools of art was announced in the press at the end of 1957, something of a storm had broken out. A public correspondence over distinguished signatures* pointed out the different systems of teaching at the two schools (the Regent Street Polytechnic employed full-time teachers while Chelsea depended largely on practising painters and sculptors of distinction teaching part-time); criticized the reduction in the total number of student places that would result (from 300 to around 250) and the general policy of concentration of which these proposals formed part; and regretted the mutual loss to the Regent Street School and to the rest of the Polytechnic, with its schools of architecture and photography, of a common physical context and the consequent exchange of ideas.

From all this, however, emerged the idea of a special Advisory Committee for the new Chelsea School of Art which was to prove of the utmost assistance to the designers of the building. By April 1958 the Governors of the two existing Polytechnics had satisfied themselves of the general desirability of amalgamation and three months later the first meeting of the Advisory Committee was held. The committee consisted of 20 people representing and nominated by the two schools and the LCC, of whom six—Sir Hugh Casson, Sir William Coldstream, James Fitton, Ashley Havinden, Henry Moore and John Piper—were appointed by the three parties jointly. The committee's duties were to examine the preliminary plans for the new school, to obtain further views on the detailed planning, to examine candidates for the post of principal and, eventually, to act as the school's governing body. One of its first acts was to set up a Building Sub-Committee of six† to discuss the detailed planning of the building more fully with the LCC architects.

On their side, the four groups in the Architect's Department of the LCC were now all the time building up the project as a whole for its next stage: its submission as a worked out, though still preliminary, scheme, with estimate of cost, to the LCC as Education Authority. This period of development lasted for some fourteen months. It was by no means all plain sailing. Although the group leaders and their teams were working in the same room, there was a tendency in the later design stages for individuals to become secretive about their own sections of the job. At one stage, the hostel block was right out on its own and it looked as though the teams entrenched on either side of Manresa Road were digging in preparatory to a pitched battle of aesthetics. It then became Powell's responsibility to restore unity. On the pretext of 'materials and finishes' he would call the group leaders and their

* One such letter was signed by Augustus John, Mathew Smith, Henry Moore, John Piper and Ceri Richards.

† This included Sir Hugh Casson, Sir William Coldstream and Mr Henry Moore.

assistants together and tell them 'by the end of the day we just must have this sorted out'. There were prolonged discussions with Chelsea Borough Council and the school over car parking provisions. Space could be made below the art school for 45 cars—at around £1,000 per car. Chelsea 'fought like tigers' for increased provision to be made, pointing out that large numbers of part-time staff were employed and that there would be no spaces for students. To have gone down to sub-basement level in order to make a second parking deck would, however, have pushed the cost per car up to possibly £2,000 and this the budget just would not stand. Again, it was planned that the equipment and services to the chemistry block should be to advanced or university standard, rather than to the normal further educational or technical college standard on which the cost limit was based. With the utmost ingenuity, it seemed unlikely that the cost limit could be maintained here. This was primarily a battle between the CAT and the Ministry. The College would put its proposals to the Ministry; the Ministry would say 'Go ahead—you can do it within the usual limits'; the architects would point out that 12-ft. ceilings would be necessary just to get the equipment in (thereby increasing costs); the Ministry administrators would call in their Inspectors—who, as scientists talking to scientists, in this upheld and even increased the standards asked for ('12-ft. ceilings? You need 14-ft.'). This particular financial thrash-around took six months of hard work before the brief to the architects was cleared.* In the meantime design work had to go on notwithstanding. It will be appreciated that, though still 'preliminary', design at this stage nonetheless entails much detailed work, so that the quantity surveyor can take off draft bills to produce an estimate of cost that will really stand up to expert and careful scrutiny. The LCC, not least its Finance Committee, and also the Ministry of Education, would have to be satisfied that this complex scheme, unique in their experience, did not exceed the cost limit or that, if it did so in certain aspects, the excesses could be fully justified.

It may be convenient, at this stage, briefly to consider the planning and design of the five main elements of the whole group separately. Their interrelationship, and the success or failure of the total design solution, may then become clearer.

The buildings to the west of Manresa Road posed no unusual demands. There were problems of access and problems of getting the volume of accommodation on to the site within the barriers of height and cost, but the building types as such are unexceptional, almost commonplace. Essentially a cellular building, the eleven-storey hall of residence—Lightfoot Hall—comprises a large number of repetitive units (the study-bedrooms) rather similar

* In retrospect an ironic battle, since from the autumn of 1966 the College became part of London University.

in scale to that of a block of flats ('egg-crate stuff'). It has loadbearing cross walls supporting hollow pot floors, and the purpose of the building was expressed by exposing the frame and recessing the window panels so as to indicate its plan and function. As forecast in the original sketch, the ground floor has been left fairly open, so that it is possible to see through it into the communal area; the entrance is on split levels, providing access both from Manresa Road and from the sunken paved area under the building. Two passenger lifts and two staircases provide circulation to the upper floors, which are planned with central corridors. Student union offices, assistant warden's, bursar's and other domestic staff offices, a sick bay, trunk stores and, of course, bathrooms and lavatories, are within the building. The key, however, clearly lay in the single unit, or study-bedroom, to be repeated 186 times (14 less than the target figure). It is in just such a building as this that architecture most closely approaches industrial design for quantity production, and that a concentration of design time upon detailing the unit concerned becomes almost obligatory. The problem here was familiar. Somehow a judgement had to be made between the relative advantages and weaknesses of a long room, with a narrow frontage, as opposed to those of a squarer space. The various areas for work, storage, sleeping and so on, define themselves more easily in the former; on the other hand, it is dark at the back of the room, away from the window, and it is less flexible in use: the bed can only be put here, the desk there, the wardrobe against that wall. Since all fitments were to be built in and would therefore have to be specially designed (by the Furniture and Display Section of Michael Powell's Division) it was vital that these decisions be taken at an early stage. In fact, two full-sized, completely furnished, mock-up study-bedrooms were built for inspection by, comment from and finally the approval of the LCC Committees, Governors, Principal and others concerned. 'It was', says Powell, 'a useful exercise, well worth the work and expense.' For Chelsea the longer, narrower room really had to be used; once this basic unit had been agreed, it became possible to plan the hall in greater detail.

Used in association with the hall of residence are the dining and commonroom facilities provided in the two-storey communal block. On the ground floor of this building is a kitchen for 800 meals, dining space to provide meals for all college students and an assembly hall. The first floor contains student and staff commonrooms; in the basement is the oil-fired heating plant to serve the whole development. In contrast to the hostel block, this is a building of large spaces and big spans, and was kept low as a visual foil to the eleven-storey slab. Design of the service areas, kitchens, wash-up, boiler house and so on, proved particularly complex on so small a site, the need for access from the King's Road creating an additional problem.

Across Manresa Road, on the east side, requirements became infinitely more specialized. In the case of the fire station, they were precise, functional, well proved in practice and unlikely to change. In the case of the art school, some of the most important of all—for example the quality of light to be admitted—could never be more than subjective. In the case of the chemistry block, they were objective enough but related to matters that are constantly and rapidly developing, so that vital changes in equipment and services, unforeseeable at the first briefing, become essential five or six years later.

The fire station was to be non-residential. The three-storey building at the corner of King's Road and Dovehouse Street contains control room and administration, together with dormitories, dining and recreation rooms. The District Surveyor's suite of offices is included here, with its own separate entrance from Dovehouse Street. Sliding poles for quick access to the four appliances are arranged at all floor levels in the administration building. A covered 'wash down' at the rear of the appliance room leads to the red tiled drill yard and tower, the yard having direct access to Dovehouse Street. The tower is inevitably a dominating feature of the scheme, of a height comparable with, but unrelated to, the hostel slab. It could not be played down and the architects sought therefore to give it a strong sculptural emphasis. The Fire Brigade were completely co-operative. Provided their detailed requirements were exactly met, they were happy to leave appearances to the architect and a completely new kind of tower—as it were, a square spiral—was the result. Below the tower is an underground smoke tunnel, with special heat equipment and extractor duct, for training in smoke and heat conditions.

The chemistry block, a three-storey building of great complexity, entailed constant and detailed discussion with the Principal and, in particular, with the head of the chemistry department. It contains laboratories, demonstration and research rooms for advanced studies in various branches of chemistry and two lecture theatres, one equipped for film projection. The noise of the King's Road, and the diverting excitements of the fire station drill yard adjoining, were major factors in the decision to turn the building inwards, round an internal quadrangle, with outside windows kept to a minimum. Many of the laboratories face on to this quiet area at the centre, the courtyard giving in addition the necessary daylighting and external wall area for natural and mechanical ventilation. From a design point of view the biggest problem here was the ducting. Whereas all flues would normally be funnelled into one main vent, it was here necessary, as part of the university standards requested, to design into the building no fewer than 17 separate flues. This in turn made for an unwelcome rigidity in the internal disposition of facilities and services. The flexibility that is so desirable in a rapidly developing field would have cost more and had therefore to be jettisoned.

Most intriguing of these specialized buildings to the east of Manresa Road was perhaps the school of art. Both the Royal College of Art and the Central School of Arts and Crafts were at this time about to move into new buildings, that for the former being of particular distinction. Chelsea, however, was the first school for very many years to require a building devoted primarily to the teaching of painting and sculpture: in this the part played by the Advisory Committee was decisive and, in Michael Powell's words, inspiring. The committee, and its sub-committee of painter, sculptor, architect and administrator, 'knew what it wanted and was completely coherent in briefing us. We had, I think, three full meetings. After that the chairman sat back and said: "It's now up to the architect. It's all his." Splendid and rather rare.'

Nor was this the only source of experience called upon. During 1958 the job architect concerned, Mr R. W. Robson-Smith, now Assistant Schools Architect to the GLC, was enabled to prepare a thesis study which entailed many weeks of research, and entailed consultation not only with members of the Advisory Committee but the Principals and Staff of eleven other London art schools. The study brought together, possibly for the first time, the functional requirements of all the main areas of teaching, from the operating height needed for Jacquard looms to the sound insulation of a lithographic stone and plate polishing room; from the average area required for a pottery student (four to six times that of other departments) to the working conditions of models. At Camberwell, it seemed, the librarian retains pot plants for borrowing by students in the same way as books. Sir William Coldstream and Mr Lawrence Gowing* were keen advocates of the lecture room, with stepped seating and full projection facilities. Above all, however, it was the spaces to be used for painting and sculpture, their lighting and, to a lesser degree, their heating, which called for the most research.

An oblong room with traditional north light along one long wall, it was concluded, restricts internal layout and use. It forces the model to be placed either in a corner or against the centre of the opposite wall—either of which entails a fan of working positions with very similar viewpoints and, at the edges, working against the light. The job architect evolved instead a north-light roof light spanning the width of an almost square studio, with a system of controllable louvres by which the degree and quality of light, both natural and artificial, could be accurately adjusted. This was tested and proved satisfactory by a small-scale cardboard model, within which a group of objects was photographed under varying conditions. Such a studio permits the model to be placed anywhere within it, with infinitely greater choice of working positions, in none of which will the student be working in his own light.

Thus it was that the Chelsea School of Art came to be designed as a sort of sandwich, with lecture rooms, administration, library and ordinary teaching rooms on three floors between doubled-height painting studios at the top of the tower and a double-height ground floor. One painting studio has the traditional north light arrangement (for still life); the other four have the

* Later to become first Principal of the new school.

louvred top-lights described (and which give to the building its characteristic sawtooth skyline). The entire walls of these studios are faced with pin-up material, without other windows or cross views, though a viewing platform with a splendid outlook has been provided, should any student feel shut in. The sculpture studios were sited at ground level to make outdoor teaching possible and to ease the delivery and storage of heavy loads of stone. The sculpture department thus forms a single-storey wing, mainly top-lit, to one side of the tower. Another single-storey wing on the other side of the tower accommodates common-rooms, kitchen and schoolkeeper's flat, and forms an enclosing forecourt, open to the street and leading to the main entrance in the double-height ground floor of the tower. Flying up the heart of the building from the latter is a double cantilevered folded slab staircase with heavy timber treads and handrails. Around this are an exhibition space and the dining-room, leading on to a further courtyard that can be used for outdoor sculpture display and teaching.

The various elements of the complex were being shaped and refined into this, their final form, over a period of three years. The sketch scheme had been presented, it will be recalled, in March 1958. The preliminary schemes and estimates were presented to the Education, Fire Brigade and Establishment Committees of the LCC in June 1959, the estimates totalling £829,400 including £85,800 for professional staff charges, fees and preparatory expenses. Further consultations with Chelsea Borough Council and the Royal Fine Art Commission having been favourably concluded by October 1959, there followed a further eighteen months' work on the 'final' scheme. Working drawings were detailed in very close consultation with the bodies and staff concerned and, together with bills of quantities, were finally completed by April 1961.

Tenders were invited from ten firms; the lowest received was from Tersons Ltd. With the addition of site layout works, items supplied separately from the main contract, and professional staff charges and fees, the total cost came out at £1,154,400. The increase over the preliminary estimate was mainly due to rising costs but the Ministry of Education nevertheless required economies to be made before agreeing to the acceptance of the tender. Agreement was eventually reached on savings amounting to £18,750, to be achieved during the course of the contract, and work on site commenced in January 1961. The contract completion date was February 1964. Programme targets exert their own influence on a building, more especially the quality of its finishes and detailing. Once construction slips behind schedule, by reason of strikes or bad weather or inefficient leadership or defaulting subcontractors, it may well never pick up again. The architect may feel like complaining that the work is not up to specification but he knows that the contractor's response will be to demand release from the contract completion date. It would take perhaps an extra four months to do the work again and, desirable or not, this is just not possible because the building *has* to be in use by a certain date: the chain reaction that would result from putting it back is not to be contemplated. Thus it was with some of the buildings in the Chelsea group. Nonetheless delays occurred, largely as a result of the very severe and exceptionally prolonged winter of 1962–3, and in the event actual completion had to be spread over a considerable period. The CAT communal block came into use in the autumn of 1963; the hostel in May 1964; the art school in the autumn of that year and the fire station not until the summer of 1965.

It had been clear at the outset that the site layout would have to be informal in character and could only succeed visually by exploiting the contrasting forms of the individual buildings. Conversely, these threatened by their very diversity to destroy the sense of urbanity that the district called for. Every effort was therefore made by Michael Powell and his four teams to keep the elevations—in particular to King's Road—as dignified and as unified as possible by an overall consistency of scale and materials. In this they were undoubtedly successful.

The construction of all buildings is of reinforced concrete, poured *in situ* except for the precast portal frames of the single-storey sculpture wing of the art school. The hall of residence, as has been said, has loadbearing crosswalls supporting hollow pot floors; the art school is of column and balanced cantilever construction with hollow pot floor slabs; all other buildings are of simple column and beam construction. The cellular character of the hall of residence is echoed, though of necessity on a different scale, in the chemistry and fire station blocks, both of which are of equal height. The drill yard of the latter, which is not an amenity open space but a paved and partly covered work area, is walled in from Dovehouse Street; this wall of dark brick is wrapped round the whole scheme, forming a dark link from end to end, from Dovehouse Street to Manresa Road and beyond. The two low wings of the art school are in the same dark brick, which thus forms a visual plinth to the entire group.

Facing materials were kept to a minimum. Where concrete is exposed, as when the frame projects beyond the window panels, it is painted with a slightly textured grey paint. The other facing material is white glazed mosaic, used on flank walls or under-sill panels. Under-window panels are in wired glass; windows and other metalwork black. The colour scheme externally is, therefore, black, white and grey with concrete, mosaic and glass the materials. Internally, bright colours have been used at entrances and in common-rooms but the predominating colour is white: furnishings, fittings, pin-ups and students provide the colour.

Those who use buildings, and pass them by in the street, for the most part assume—if they think about them at all—that they embody an architect's ideal. But where today in architecture exists the possibility of absolutes of excellence? We have seen some of the hidden factors, lying quite outside the functional demands of these buildings, which helped to shape their qualities: a big city's pressure on space, the wishes of powerful neighbours, cost limitations, height restrictions, a bad winter. Within the given factors operating at the time, these buildings may be considered more than usually successful. They form a convincing urban group: well mannered, thoroughly worked out and, to the east of Manresa Road, lively even. Indeed, the school of art received a Civic Trust commendation in 1964 for its contribution to the environment. Yet whose was the creative response to all the problems of the brief which resulted in so satisfactory a solution? The answer must be: no single person. The buildings represent the accumulated decisions of successive designers in a self-renewing team. If ever proof were needed of the resilience of the group as an organism tenacious of life, it would seem to be evidenced here in Chelsea. For the extraordinary fact is that, due to promotion, switching within the Department and wastage through resignation, not one single member of the team, other than Michael Powell himself, saw the job through from start to finish.

8. The Built Environment: II

At number forty-nine in the via Dezza, in Milan, stands what was once a large garage. Since 1950 it has housed the remarkable studio of Ponti-Fornaroli-Rosselli. It is an unusual place by any standard, and something of a magnet to architects and designers visiting Italy. Its continuous, top-lit space is divided only by light partitions; within it work the three partners who give the firm its name and their thirty or so colleagues: architects, designers, engineers, surveyors, secretaries.

The PFR Studio reflects, on the one hand, the personalities of its principals; on the other a group personality which is itself in part their creation. It is no accident that the partners are sharply contrasted in age and interests. Gio Ponti, a born impresario of the plastic arts, first set up in practice as an architect in Milan in 1925. From the beginning he regarded the fine arts and fine craftsmanship as significant fields of activity for the architect. The essential nature of the group took shape when, five years later, Ponti was commissioned to undertake the Montecatini building and formed an enduring link with the young Milanese engineer Antonio Fornaroli in order to strengthen the technical design resources of the studio. Since then the concept of collaboration has grown clearer, the group has grown closer and the overall creative output of its members has increased, without compromise, they say, to their varying characters and personalities.

Characteristic of the group today is the way it spans the generations. In Ponti, Fornaroli and Rosselli we see three successive generations, three different stages of experience in architecture, in life. Thence springs a particular dialectical working relationship within the studio, in which three levels of maturity, of design approach, of critical judgement have to be resolved. The age differences between the partners coincide with differences in their cultural backgrounds. Ponti's early interests lay in painting and handicrafts. Fornaroli, with his background of civil engineering, from the beginning was concerned to investigate the then uncertain relationships between technology and architecture. Rosselli's early architectural experience is complemented by his interest in industrial design and the search for methodology in the design process.

The teamwork of these three reaches different states of equilibrium according to the nature of the work in hand, acquiring a greater or lesser personal emphasis in keeping with the weight of the contribution made by each. The virtues of such flexibility have been amply borne out by the very wide range of work carried out since 1945. Even when one of the members is given a free hand on a job, it does not mean, says Rosselli, that collaboration is ignored; the effect, more often than not, is to make it more lively. To PFR collaboration means individual initiative plus group control; its purpose is still to strengthen the creative capacity and technical efficiency of the practice as a whole.

When Ponti left his via Saffi office (where he had also lived) in 1950 to set up the present studio, he had three things in mind. First, he wished to unite in a common atmosphere all the activities connected with design: not only the technical and design facilities required by an architectural practice, but furniture design, industrial design, graphic and editorial work as well. Secondly, he hoped to make this many-faceted organization a practical training ground for young people working in those fields. He aimed, thirdly, to make the studio something of a display centre for Italian art and industry; even, as a place of experiment and research, something of a 'live' exhibition in itself. Fashionable elegance, a prestige front, were not in his thoughts. He sought only a big and comfortable place to work in, free of the limitations imposed by traditional offices. It is an indication of his aspirations that, from the first, partners, architects, designers and the studio's secretary worked cheek by jowl together.

In his general aims he was to a large extent successful. The via Dezza practice has tackled a very wide range of work and undoubtedly benefited from the fact that its own intelligence and propaganda unit—the editorial offices of *Domus*—was in one part of the studio. Under the pressure of their commissions, however, the group's field of activity gradually shifted to the planning and execution of complex, large-scale jobs. This readjustment was occasioned primarily by the growing scale and complexity of architecture itself, which began to demand of the group a more methodical approach, a tighter degree of organization, a much broader kind of collaboration: in the words of Rosselli, 'a new professional attitude'.

The total strength of the studio has varied between twenty and thirty. Today it consists of the three partners, ten to fifteen architects, two to three engineers, two to three surveyors, a special secretary and an ordinary secretarial staff of four or five. At any given moment there is usually a handful of foreign architects and students participating in some capacity or other in the work of the studio. Over fifteen years, more than fifty such, of many nationalities, have taken their place with PFR. 'Their presence', says Rosselli, 'has helped to establish a sense of co-operation; their liveliness and ability have acted as a stimulus to original research.'

Design control is exercised by the partners; collaboration between them and the *ad hoc* groups into which the staff are formed as the work may require is always based upon a programme drawn up by the partners for each particular commission. Responsibility for architectural decisions is shared between Ponti and Rosselli, who have never veered from the principle that this prerogative is properly theirs. Responsibility in PFR means not only the provision of the fundamental idea for a project but the full development of all its technical relationships and correlations. In this belief that the 'creative' aspects of design are not to be divided from the overall technical aspects, the size of the practice has been limited to a number that permits the architects to do the designing themselves and remain completely responsible for a commission from beginning to end.

On the other hand a number of other groups are closely associated with the PFR studio, although not directly part of it. For the calculation of reinforced concrete, for example, a close and constant collaboration has been maintained in recent years with Professor Locatelli, chairman of the department of the Science of Construction at the Politecnico of Milan. Professor Locatelli has calculated and designed the structures of all the most important steel and concrete buildings commissioned from the studio, from the Pirelli building to that for the *Corriere della Sera.* There have also been fundamental exercises in collaboration with other engineers, most notably with Nervi and Danusso. Important contacts have been made with foreign consultants: with quantity surveyors and landscape architects in England, for example; with American specialists in the design of certain mechanical plant.

Around the central design nucleus of the PFR studio itself, that is to say, exists a series of specialized contacts and groups whose participation is sought and proportioned according to the nature of the work in hand. This is common professional practice in Italy, as elsewhere. In the case of PFR one senses a sharper ideological desire to win the collaboration of superior technical talent in dealing with themes of importance; to bring about, inside the studio and out, as the prerequisite to genuine research, a meeting of minds already passionately concerned with higher standards in their own field.

The distance of these consultants one from another, say PFR, is only a minor inconvenience, easily overcome by modern communications, by the methodical formulation of problems and by the need to design by ever more specialized systems and techniques. Co-ordinating the contributions of such offices turns essentially on the architect and his method of designing. PFR aim to emphasize and perfect the capacity for synthesis rather than analysis, to make things converge and combine rather than to break them up into specialized fragments. There is intense initial concentration upon the plan by architect and consultants; the work is envisaged at all times as a complex unity; all its future environmental and technological implications are as far as possible anticipated. This means that successive syntheses are like concentric circles round the initial idea; it also requires that the initial idea should hold within itself the germ of harmonious future growth.

It was the Montecatini building which really established the group's approach to architecture, every problem and sub-problem of which, they recognized, required research, planning and organization. Well into the post-war years Montecatini still represented, at least in Italy, an historic precedent in the use of new materials, new technical services such as air-conditioning, and the rational organization of the spaces required by the administrative activities of an industrial society.

With 1955 came the first studies for the Pirelli building, on which the work of the PFR studio was to be concentrated for some six years. The technical problems of the slab and point block are now no longer novel, their solution is familiar to every fourth-year student. Pirelli was the first true high-rise building in Italy, however, and its scale demanded of its designers a much wider range of collaboration, larger teams of technicians and greater co-ordination with industry, than previously. The same period saw the construction, both in Italy and elsewhere in Europe, of other buildings of a comparable size and complexity. The situation thus developed into a sort of international competition in ideas. The Pirelli project was indeed first approached with ideas; methodology and organization followed to aid the formulation of the fundamental concept, its development and its final constructional expression. As we have seen, the designers sought not to impose preconceptions upon the form of the building but to allow this to be shaped by its own internal logic. They listened to the building; they sensed its imperatives and sought to comply with them.

This was their aim once more when, at the end of July 1961, they were invited to plan and construct a complete government secretariat in the new capital city of Islamabad.

Islamabad stands within a crescent of hills, some ten miles to the north-east of Rawalpindi. Here historic trade routes meet: one the Grand Trunk Road of the British, from the Khyber and Peshawar to Lahore and the south-east; the other north-west from Rawalpindi to Kashmir. From these foothills of the Western Himalayas, 1,750 feet and more above sea level, the Potwar plateau falls gradually to the south. It is gently undulating country, extensively eroded by the monsoons, which have left *nalas*, or gullies, up to fifty feet deep, and stripped of its vegetation over considerable areas by the herds of nomadic tribes. Only four years after the ground was broken, Islamabad's population was 30,000. The last section of the last Ministry had been shifted from Karachi. Housing, shops, schools, mosques and hospitals were in being: spick and span and well cared for. An atomic reactor was under construction. The river Rawal had been dammed to form the first of several great lakes in the national park. Trees were everywhere, in spinneys, groves and avenues, a foretaste of the forests and gardens promised for the bare hills in the background.

'Artificial' capitals, deliberately created as such, are not uncommon; but neither are they numerous. The deliberate act of national will that brings into being a Washington, a Canberra, a New Delhi, a Brasilia, results only from the most pressing needs for decentralization, for the vitalization of underdeveloped terrain, for the removal of government machinery from the frenetic activity of existing commercial centres, for a new symbol of national prestige. For Pakistan the overwhelming need was a city that would embody the country's national aspirations and put an end to what has been called its 'transit camp mentality'.

Pakistan came into existence as a result of the partition of British India in 1947—a nation of two territories 1,100 miles apart, lacking both a sense of identity and a traditional seat of government around which a sense of nationhood might grow. Karachi, the interim capital, is hot, humid and short of water. Possible alternative sites were canvassed by successive governments but it was only with the accession to the presidency in 1958 of Mohammed Ayub Khan that the movement for the creation of a new capital gained purpose and urgency. A site selection committee was appointed in February 1959. In four months (surely the fastest action of its kind on record!) its recommendations had been made and accepted by the government. In the East, Dacca was designated 'second' capital and seat of the legislature; in the West, the 'first' capital and seat of the executive was to be, though the site was as yet nameless, Islamabad.

The whole vast project was now pushed ahead with all possible vigour. In September of the same year a Federal Capital Commission was set up to prepare a master plan and programme. Its preliminary plan was approved in May 1960; the final master plan in October. Six weeks earlier the Capital Development Authority had been created as executive agency to implement the plan. Its first Chairman, Mr W. A. Sheikh, exhibited drive and enthusiasm. Over 350 square miles were pegged out; 6,000 families were moved and resettled elsewhere, their villages abandoned. Twelve months later, in October 1961, work began on the ground. By October 1963 the first civil servants arrived and took up residence in their new homes.

Nor was speed the concomitant of inadequate preparation. The staffs of the Federal Capital Commission, and its successor the Capital Development Authority, were augmented by British architects and planners, engineers and other specialists under the Colombo Plan of aid and technical assistance to Commonwealth countries. And, although responsibility for much of the detailed design and execution was assumed by the Authority, the major planning and architectural decisions were entrusted to international consultants. To the Greek Dr Constantine Doxiadis was given the task of evolving the master plan. Fourteen committees of 100 experts, drawn from 46 Departments of Government, carried out the preliminary studies, surveys and research. These studies covered such matters as site and climate, traditional town and village patterns, the basic concept of the city, primary traffic movements, the sociological and economic structure of communities, local materials and methods of construction.*

The plan which was based on these reports reflected Doxiadis's concept of the 'open-ended' linear city: capable of infinite extension, with other areas of the city never more than a relatively short distance from the spine of the central core. The main highways define three principal divisions: Islamabad, Rawalpindi and the great national park area of nearly 60,000 acres. They also impose a grid pattern upon the sectors into which the main metropolitan areas are divided. Elaborate studies were carried out with the object of establishing zoning regulations by which the main growth of the city could be controlled. The result is a complex hierarchical structure in which the size of roads, the facilities and amenities (including schools, shopping and open space) for five categories of community, permissible heights for buildings and many other factors, are precisely laid down. For the civic and commercial centres, the Capital Development Authority has itself prepared plans and elevations which all private developers must follow, with minor modifications, and its designs for various housing types, primarily intended for official staffs, are also made available to the general public at nominal cost.

In contradistinction to the diagrammatic, abstract nature of this part of the plan, liveliness and variety were sought in the administrative sector which, at the apex of the triangle forming Islamabad, set dramatically against the rocky escarpment of the Margala Hills, would contain the main government buildings—the emotional core of the city. With this object, the bold decision was taken to assemble an international team of architects for the purpose—unlike, it may be noted, Chandigarh, which was entrusted primarily to le Corbusier, or Brasilia which was entrusted to Niemeyer. 'After careful deliberation, we decided that we should not follow this method but adopt a novel approach of achieving unity through the concerted efforts of a group of world-famous architects whose work was basically in sympathy. It was felt that by this method, we would avoid monotony, and the repetition of architectural clichés. On the other hand, we would have at our disposal the joint technical resources of a group of renowned architects who would be able to impart a certain variety and vigour to the architectural conception, without destroying the essential unity.'†

Thus it was that the Ponti-Fornaroli-Rosselli studio received an invitation, on the last day of July 1961, to participate in the creation of Islamabad. Others to be brought into the team included Robert Matthew, Johnson-Marshall and Partners, to be consultant architects for the Administrative

* A summary of these reports is included in 'Islamabad, the New Capital of Pakistan', Doxiadis Associates, *Ekistics,* November 1964.

† 'Islamabad. New Dimensions in Architecture', by Khawaja Zahir-ud-Deen, Director of Planning. Capital Development Authority, Rawalpindi, July 1964.

Sector; they, with Arne Jacobsen, Louis Kahn and Gio Ponti, were to be responsible for the more important buildings in this sector; Derek Lovejoy and Associates were retained as landscape consultants. The main task foreseen for the PFR studio was the design of a large secretariat group that would house the Pakistani civil service. The nature of the brief was set out thus:

> The master plan for Islamabad provides for the construction of most of the monumental building including the National Museum, the National Library, etc., in one compact zone described as the 'Administrative Sector'. A challenging feature of this sector is its location and its commanding position in relation to the rest of the city.
>
> In developing the Administrative Sector the aim of the Authority is to create a civic centre which is functionally sound, economic, suitable to the physical and social environment of Islamabad, and architecturally significant. The individual buildings should be contemporary in design, capable of serving the present and future requirements of a growing country such as Pakistan, and also reminiscent of the great tradition of Muslims in architecture. Each building must have an identity of its own, and at the same time should be so related to others that the entire complex has an overall unity and character. The intention of the Authority in placing emphasis on tradition is to re-establish a creative connection with the past and not to imitate the traditional forms blindly.

The Ministries to be catered for were listed: twelve in the first phase, nine more in the second phase, with separate provision to be made for the Cabinet Division, External Affairs and Defence. Altogether it was estimated that some 600,000 square feet would be required. In addition, it was proposed that PFR should design for the Secretariat area a library, lecture hall, mosque, telephone exchange, fire station, guard room, post office, cafeteria and sundry garages, car parks, cycle stands, etc. For some of the buildings special requirements were tabled: strongrooms, fire-proof archives and so on; for all, a relevant seismic factor was to be allowed for in the calculations for concrete. The cost of the buildings was to be kept within the estimate of 30 million rupees (£$2\frac{1}{4}$ million); their construction was to be completed within two years, as it would not be possible to move Government Departments from Karachi in ones and twos but only *en masse* in the phases foreseen in the overall programme.

This was a sizeable challenge in view of the imponderables inevitably associated with work in another continent, where many of the skills, experience and resources which are taken for granted by Western technology could not be expected to be found. The task divided itself into three natural phases. There was first the functional and visual co-ordination of the buildings with the demands of the master plan; there was the detailed planning and design of the buildings themselves; there was thirdly the actual construction

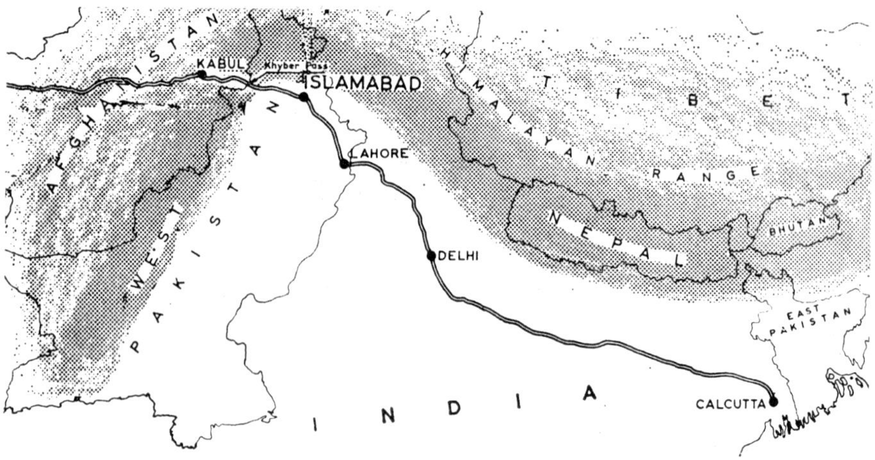

After the partition of British India, it became essential for the new state of Pakistan to create a capital city. The choice of site lay between Karachi, hot, humid and low-lying, and the Rawalpindi area in the foothills of the Himalayas. A special Commission recommended the latter, on the old Grand Trunk Route that runs from Kabul to Calcutta. Here should arise the new city of Islamabad, above.

Below, the Doxiadis master plan for Islamabad.

Centre, top, cupped by the Margala Hills, is the administrative quarter of Embassies and public buildings. F6, F7, G6 and G7 are the first four housing areas to be developed. From here the city will grow southwards and westwards, along two spines of civic and commercial areas, until it is linked with Rawalpindi, ten miles distant. At the apex of the triangle, within the administrative quarter, the plan placed the Secretariat building to house the Pakistani Civil Service.

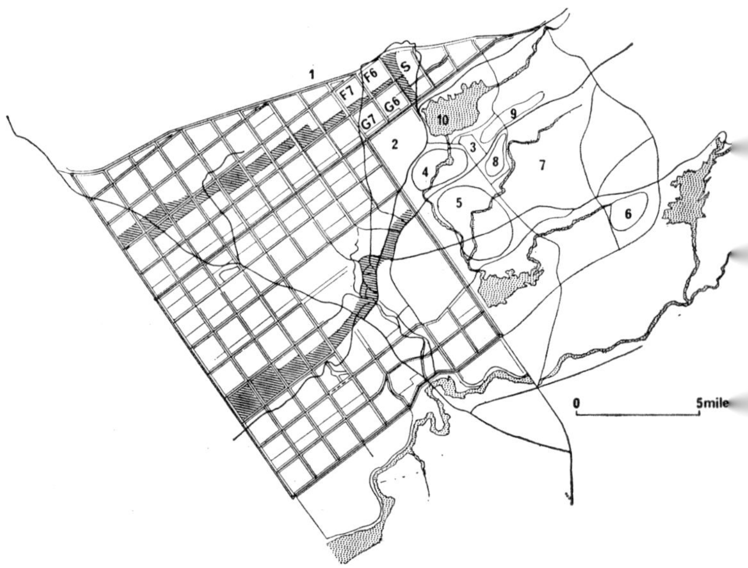

Key
1. site for national mosque
2. Shakaparian village and hills
3. medical centre
4. zoo and botanical gardens
5. exhibition area
6. atomic research centre
7. national park
8. educational institutions
9. national university
10. Rawal Lake
F, G. housing sectors
S. secretariat

Above, the first volumetric model prepared by the Ponti-Fornaroli-Rosselli studio. Its object: to establish a valid, viable approach to the problem. With the agreement of Sir Robert Matthew, the co-ordinating architect for the administrative quarter, they were seeking a solution that would be more flexible in use than that proposed by Doxiadis (who is to say what may be the requirements of a young Government a few years from now?) and which would sit happily against its magnificent backdrop of the Margala Hills. Below, part of a subsequent and a more detailed site layout.

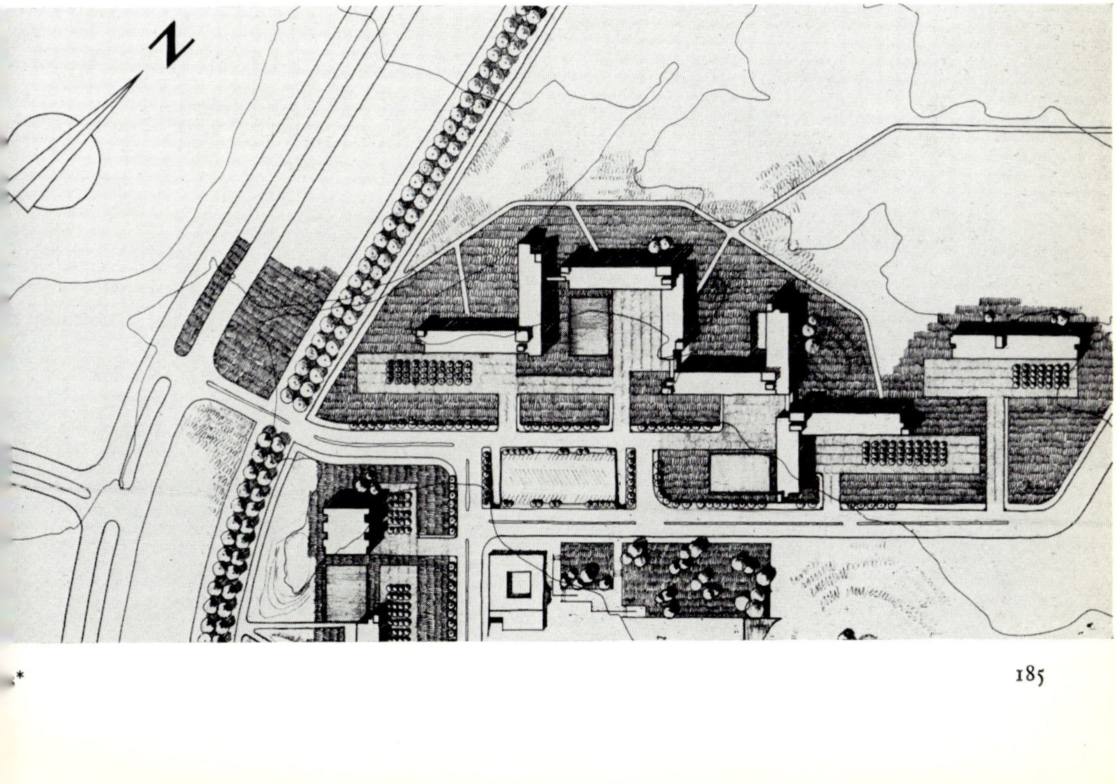

A final model was now constructed, above, both pages; this embodied substantially all the main characteristics of the group as built. For long before this, however, studies had been in progress to determine the degree of protection from the sun that would be afforded by different elevational treatments. It was PFR's aim to evolve a number of different precast elements—varied according to their orientation and the degree of protection required—that could be inserted into a basic structural system to be employed throughout the whole

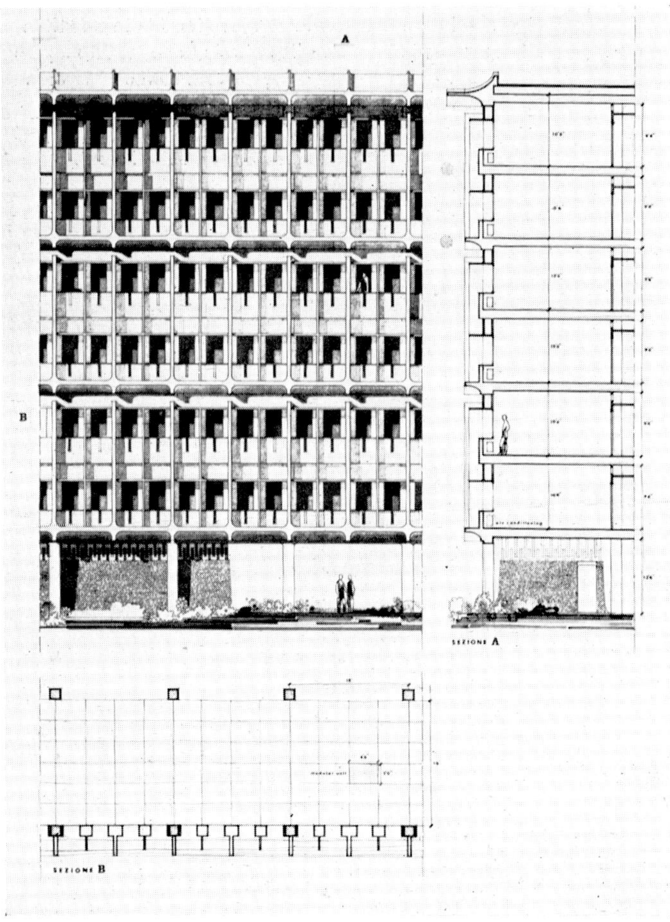

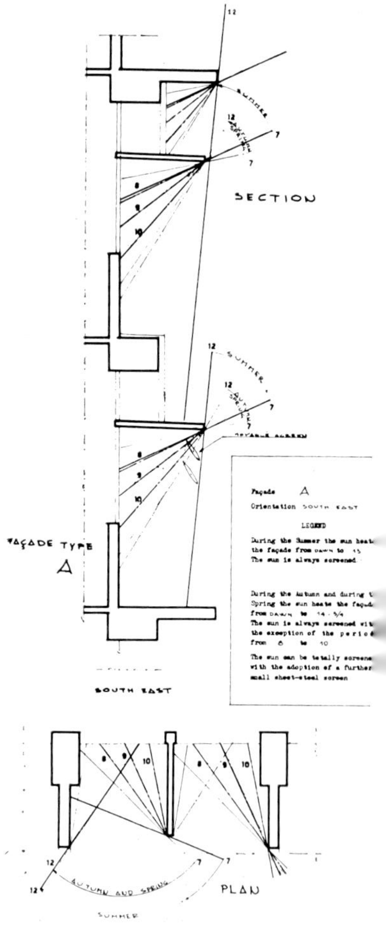

oup of buildings. Simplicity was essential, for skills ken for granted in the West were not to be expected Islamabad; even so, problems of fabrication and ection arose: the living and working conditions on site may be glimpsed below. The completed elevation at the foot of the page, its perforations and general articulation giving it a faintly oriental character, may be compared with the earlier drawing on the facing page.

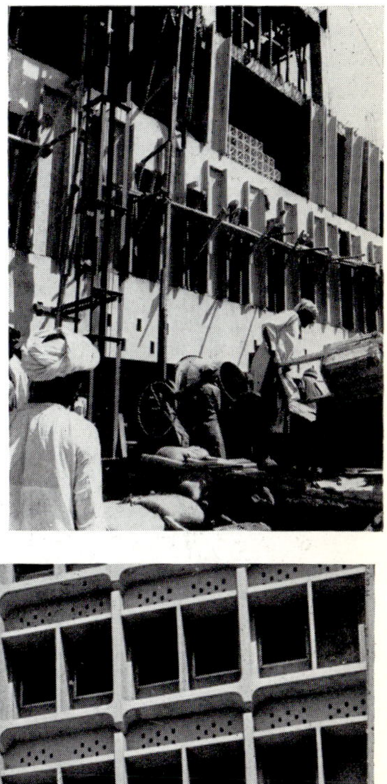

Two other elevational treatments, that on the left consisting of simple vertical screens for East and West façades, that on the right for elevations more exposed to the sun. Here, too, may be seen the careful detailing of surrounding hard-surface areas. Landscaping—in the hands of the English firm of Derek Lovejoy and Associates—has been one of the most marked features of the planning and construction of Islamabad—at all levels. The completed housing areas are already green. Extensive planting has been undertaken to clothe the Margala Hills, against which the Secretariat buildings, seen from the South (below), raise their towers—contemporary echoes of the domes and minarets of the traditional Islamic city.

of the buildings, together with the associated problems of long-range communication and the best utilization of locally available skills and resources.* These were not new problems to PFR. The studio has undertaken jobs in places as far apart as Caracas and Baghdad, Eindhoven and Hong Kong. Nonetheless, Islamabad promised to present them in acute form and, as always, the first need was for information.

PFR's initial visit to the site of the new capital coincided with the commencement of work on the ground. The first roads were being prepared; the first housing areas were under construction. The detailed planning of the administrative sector was being discussed with Sir Robert Matthew and with him, on the spot and by correspondence, over several months, PFR hammered out the general principles of the Secretariat zone. Co-operation here was smooth, the work quickly and satisfactorily completed. PFR are quick to praise Matthew's skill as a co-ordinator and the authority with which he ensured that decisions were respected by all concerned (his formidable abilities as an administrator were glimpsed in the previous chapter). A first layout plan for the Secretariat group was completed in April 1962; the final and detailed proposal was submitted in October of the same year.

As has been seen, Doxiadis' outline plan was rigidly rationalistic in its approach, imposing upon the terrain a regimented pattern of abstractions—albeit based upon the most detailed site and climate studies. Government Ministries he foresaw as ten- or fifteen-storey blocks, set in a row along the large National Avenue, against the continuous rocky escarpment of the Margala Hills. Matthew and PFR sought a less autocratic solution.

> While respectful of rationalist theory, say PFR, we were of the opinion that the country could suggest a different kind of approach, not on a monumental but on a very human scale. We thought of a group of buildings in harmony with the Pakistan environment and tradition.

Where does the idea of a work of architecture come from? How far does actual method or procedure influence it?

> We believe, says Rosselli, that it is only by setting out with a fundamental idea that method can develop and influence the architectural process to the very end. Before this, method is the investigation of those relationships and component elements necessary to the making of a choice, the formulation of this guiding principle. There is then a point of transition, between research and planning, where a particular kind of synthesis is required—the hypothesis, or 'idea'. It is here, in our view, that method can reveal the decisive characteristic, the quintessential nature, of a project. All-important at this moment are the personal resources and intuition of those involved in the project. Even a belief in method must leave a place in it for those personal syntheses which go

* For example cut-and-fill, which would normally be done by earth-moving equipment, here became the responsibility of tens of thousands of donkeys.

beyond the dead ends of pure analysis. Thereafter, method once more becomes essential.

PFR now set themselves, therefore, to evolve a methodical procedure for isolating and identifying problems connected with the project: the type of relationship existing between the natural environment and the new city, between the forms of political organization, work patterns, social organization; between these and architecture. More had to be discovered about the relationships between the economic and the industrial and technological environments, between programmes and time schedules, between the level of technical skills available for construction and the likely demands of the project itself; in brief, between reality and PFR's aspirations.

At this stage the project gave rise to many hypotheses, many alternatives. On the one hand the European tradition provided examples of historical palaces and public buildings; on the other, the modern movement offered examples of rational structures and technological development. The relevance of both, in simple terms, was limited. 'We were unwilling to transport to Pakistan the experience and ways of Western architecture but wished rather to create a new architecture that would reject both monumental and mechanical functionalism.' The nature of the Rawalpindi landscape, local culture and tradition, the climate: the subtle problems of environmental relationships were not to be burked. Indeed, did not the architects' brief make specific reference to the Muslim tradition?

Somehow, too, a suitable system had to be found which would provide flexibility in use; which would allow for the possibility of future expansion within the original concept and which would not stiffen into monumental shapes unsuited to changed ways of working. For who can say, of a nation so young, that its political structure is final; that its very Ministries may not suffer great transformations; that future development will not bring the creation of new Departments, the abolition of others?

Thus was born the idea of a unified system of work-spaces, rather than a group of single self-contained units. The Secretariat buildings took shape as two groups, each of four L-shaped blocks of five to six floors only, marked by vertical elements of towers (the vertical communications) fitted into the node points. The blocks would be accommodated at different levels, to avoid disturbance of the natural contours, and the interrupted line of the buildings would allow the skyline of the Margala Hills to enter and blend with the architecture.

Organizational problems now loomed large. PFR were asked by the chairman of the Development Authority for their maximum effort so that the timetable for the project might be cut to the minimum. It was clear that, if this request were to be met, an early resolution of all related problems at the design end was essential; that communication with the site should be smooth and unambiguous; and that technical procedures on site should be appropriate to the skills available.

To achieve the first of these PFR, themselves members now of a larger international team, would have to split the planning of the project among further groups, internal and external, specializing in architectural design, structural planning, services and installations, landscape design and so on. The co-operation of Professor Pietro Locatelli was once more sought in the calculation of reinforced concrete; a second engineer, Mario de Bernardinis, and a geologist, Professor Guglielmo Meardi, were further brought in as consultants. On electrical, air-conditioning and plumbing problems, the studio had the assistance of W. S. Atkins and Partners in the first phase; of A. Saadat and Co. in the second. At the same time consultation went on continuously with the CDA, Doxiadis, Sir Robert Matthew, Derek Lovejoy and their various colleagues. Sometimes meetings were at Rawalpindi (partners from PFR visited Islamabad nine times between 1961 and 1966), either for general meetings of all consultants, called twice a year, or to sort out site problems. Sometimes Doxiadis, Percy Johnson-Marshall, Lovejoy or a partner from Messrs Atkins would call in at the via Dezza on their way through to Rawalpindi. Sometimes partners from PFR went to London for special meetings. On some of their travels PFR were accompanied by their structural consultants, or by Professor Meardi the geologist. Correspondence between Milan and Rawalpindi was regular and frequent.

Much of the discussion regarding the buildings centred on the purely technical procedures to adopt with the materials to be employed. In the West such choice would be based on an adaptation to particular problems of a widely held corpus of experience. In Islamabad, emphasis was put by the Development Authority on the exclusive and novel character of the choice, and on its educational and programmatic importance to other and future construction in the capital. The very high technical standards of Western architectural design, impossible to realize anyway in a less highly geared economy, simply would not correspond to the local way of life, work and culture. The building techniques which were, in fact, adopted betrayed their origins and were not easy to graft on to an Asiatic country with that particular climate. The West's technology, PFR felt, was revealed as a tool designed for a given society rather than a system of universal validity.

Islamabad has an extreme annual temperature range of 70–80 degrees Fahrenheit. The winter weather is magnificent but the summer, if less overwhelming than in Karachi, is still very hot, while the monsoons bring two-thirds of the total rainfall in the three hottest months. The climate demands strong protection for the exterior surfaces of buildings, while excluding most of the cladding materials used in the West. Local materials, brick or stone, had all the emotional significance and charm of harmonizing with the setting. They were too closely bound, however, PFR felt, to the scale of traditional construction and to the artisan's pace of building, to match the structural scale envisaged for the Secretariat group. Again, past generations constructed their monuments in stone because they associated an aristocracy of materials with an equivalent scale of human values. These buildings in Islamabad, however, were not to be dedicated to the monumental, to bureaucracy, to the past, but to the human scale, the present and the future.

In the event, reinforced concrete was chosen for its overall characteristics. As a technique, its demands upon the building worker are few: little more in fact than the construction of the wooden form-work; as an operational practice it does not require a large number of specialists; it offered solutions to technical problems at several levels and great resistance to climatic and allied hazards; its cost was attractive and it was appropriate to the type of architecture envisaged. Over and above all this, however, it promised a simplification of procedures in the long-range control of construction. Where, as a result of over-specialization and inter-professional rivalry, drawings are divided into those dealing with the structure and those dealing with the 'architecture' of a building, their recomposing—the careful sewing together of the elements—poses an additional problem of liaison. It is sometimes difficult. At best it further complicates matters, requiring a greater number of drawings and comparisons. By carrying the technology of reinforced concrete to its logical conclusion, the PFR studio and its consultants sought to ensure that structure derived from the architecture and architecture from the structure. The fact that the whole, structure, walls and elevations, was designed in the same material avoided successive interventions, the overlapping of often different techniques and any possibility of ambiguity or misinterpretation. Nothing could be changed. Everything was definitive.

Architectural design is a system used by technicians to communicate with one another, a procedure for breaking down a form into its constituent elements so that it may be reconstructed completely and in all its complexity. Today, design is entrusted with most of the information which was exchanged in former centuries through personal contacts. It can synthesize the different disciplines and incorporate information from many different sources at many different levels, sometimes embracing sub-problems of great sophistication, and thus forms a complex technical code. The architect's ability to give his design all the information needed for its complete execution, while making it bear simultaneously the full import of his aesthetic aspirations, is conditioned by his method. It was vital for PFR, in working at such a distance from their consultants, their client and their contractor, to evolve a procedure in which nothing was left to chance, everything had been foreseen.

As a code, design is intended to be deciphered and read and it is on the capacity for reading designs that the operational mechanism in building yards depends today. The Islamabad project raised this problem in acute form. The project was studied in Milan in every detail, transferred to drawings of various categories, sent by air to Karachi, where they were cleared through the Customs by a CDA officer, and thence to Islamabad where the 'code' was deciphered and transmitted to the site and the building yard, there to be reconstituted in the structures and forms foreseen in Milan. This procedure was maintained for three years at the rate of 40 to 50 drawings a month: 94 despatches altogether, apart from the material brought to Rawalpindi personally by PFR partners. In all, some 1,600 drawings fuelled a site bustling with the activity of 5,000 workers, 100 technicians and 30 engineers. The architect's presence on site, that is to say, made itself felt not, as traditionally, through his physical presence but through his designs and his drawings.

The process was not without its alarums and excursions—how could it have been?—more particularly at the outset. In Milan, the very flow of drawings created a crisis in the earlier stages. The sheer number of details to examine, indications to check, constituted a phenomenon of greater than normal proportions. A small slip in the sizes, which escaped the draughtsmen, created difficulties on the site where, in the early days, there were not enough people sufficiently experienced to interpret the drawings. An engineer from CDA, a Mr Javaid, spent five months in the PFR office in Milan during the final project work, in order to become familiar with the scheme and its details and finishes. He then followed the works on site but left the Development Authority a year later. Order really came to the site with the admission of a German engineer and an English clerk of works, Mr J. Arthur Biggar, to work in co-operation with the Pakistani technicians of the CDA. Thereafter, increasing responsibility for site works was assumed by the latter until, at one stage, fifteen to twenty Pakistani engineers were taking care of the various buildings.

It had been decided to use prefabricated precast concrete elements for the façades, set in the frame of the cast concrete structure. Apart from reasons of aesthetic unity, this approach made for greater speed of construction. On his first visit, however, Rosselli found that the prototypes for these precast elements were quite inadequate; that the methods for their manufacture were very primitive and that, owing to the lack of even the most basic lifting gear, their fixing in place raised enormous difficulties. It was a bad moment. In the face of such fundamental shortcomings the gloom of the architects was extreme. They pressed on, however, discussed the difficulties and examined new proposals. The contractor's technicians and Mr Biggar between them evolved a way of manufacturing the precast elements satisfactorily. A means was then found of lifting the finished components, by cable over a pulley wheel attached high up on the building and powered by a tractor moving away from the building.

Thereafter progress was relatively smooth. Mr Biggar created order among the drawings and established a scheme of reference which permitted their easy consultation. At their end, in the via Dezza, PFR received a site report every two weeks. First excavations had taken place in April 1963; the first casting a month later. By the spring of 1966 work was nearly 90 per cent complete; and indeed, the first buildings would have been ready for occupation on schedule in October 1965 had the sub-continent not found itself that year at war.

Oddly, with completion in sight, new problems, new doubts, new anxieties became apparent, the real magnitude of which is not yet to be gauged. They make it clear, if proof be needed, that the harnessing of a creative team is more than a matter of mechanical administration.

We have noted the names which, in 1961, were associated or to be associated with Islamabad: Doxiadis, Jacobsen, Kahn, Sir Robert Matthew, Ponti, Lovejoy. By the end of 1966, the last three alone were still employed by the Development Authority. First of the architects to go was Jacobsen, whose cool, sophisticated Parliament Building failed to commend itself—perhaps on grounds of technique. Responsibility was taken over by Kahn, who over two and a half years produced three schemes for this group of buildings, none of which proved acceptable, on grounds that they did not reflect the Pakistani character (though the same architect's somewhat naïve massing of perforated cubes and cylinders is apparently acceptable in Dacca). Kahn in his turn was replaced, in November 1966, by Edward D. Stone, who has produced an Orientalized Beaux Arts scheme which presumably flatters national susceptibilities.

Nor was it possible for PFR's own buildings to avoid some trouble. The original brief, it will be recalled, made specific reference to 'contemporary design' which would also be 'reminiscent of the great tradition of Muslims in architecture'. PFR's work has been criticized by some Pakistanis as uncharacteristic of the country, and a change in the leadership of the CDA has led to an edict that future buildings must reflect the Islamic genius more directly.* A tentative intervention to give some of PFR's buildings a 'face-lift' was squashed by the chairman of the Authority himself but here, clearly, is an aspect of design on which feelings must run deep. The future of Islamabad would seem to be connected with the solution that is reached.

Design has indeed become international in many ways and the general premises on which it is based, as has been stated earlier, command very wide agreement. Taste and style, however, are still tied up, as they always will be, with personal and national value judgements. Architects commonly build

* Though there are said to be only twenty-five local architects in the whole country; by the autumn of 1966 the CDA had recruited only one qualified architect-planner, one architect and one planner of Pakistani nationality.

nowadays at great distances: le Corbusier in Japan, SOM and Jacobsen in Britain, Sir Basil Spence in Rome and so on. Embassies, hotels and exhibitions, however, which provide the staple subjects of such building, are commonly regarded as a means of 'showing the flag'. The buildings, that is to say, are naturally designed as far as possible to sit down amicably with their neighbours but are not expected to ape the national styles of the country concerned—just the reverse. The extent to which it is possible, or right, for a designer to attempt to assimilate and express a culture that is not his own is not a matter in which we have logged up much experience. It is nonetheless a problem of great significance for the developing countries.

The Capital Development Authority's brief to PFR showed a sensible grasp of the issues at stake: 'The intention of the Authority in placing emphasis on tradition is to re-establish a creative connection with the past and not to imitate the traditional forms blindly.' This distinction between the meaningful core of tradition and the academic formalism of its outer shell accorded exactly with PFR's own thinking.

> Tradition [they say] is often invoked in order to avoid the obligation towards the problem of our own age, or as a means of obtaining authority's blessing and public approval. All European countries are steeped in tradition—none more than Italy—but we understand also that a tradition which does not contribute to the current development of our living and thinking cannot be said to exist as a continuing force. As people who know and love history, we cannot use the easy escape provided by formal tradition in the academic sense. As Western architects designing for the East we tried to establish a method of approach to Muslim tradition, which we interpreted not as bounded by the forms of the past but by the general ambience of the past, by present culture, ways of living and the local climate.
>
> We went beyond rationalism in our interpretation of the environment. On the other hand, faced with problems of technology, economy and structure, we had too to retain a rationalist position. We could not ignore the fact that these buildings belong to the contemporary expression. The buildings in Islamabad are not European buildings imported there. They are completely different from those we have designed for Italy and elsewhere.

PFR were supposed to design the interiors for their Secretariat building. Furniture prototypes have been made in Rawalpindi to PFR designs and Italian furniture has also been examined. Pakistani painters and sculptors have been contacted in connection with particular parts of the Secretariat area. It will be revealing to follow the future course of these collaborations, of these buildings, of the whole administrative sector of Islamabad and of the city itself. The pace of construction slowed for a time as a result of the war with India but, having been brought to life, as undoubtedly it has, Islamabad is likely increasingly to generate its own forward impetus. As yet, it is a one-industry city, with the Secretariat as its *raison d'être*. In the early 'seventies, however, light industry on a considerable scale is scheduled to appear on the scene. Already 60,000 inhabitants are accommodated in sectors G6 and the completed parts of G7 and F7. Some of the individual buildings in these sectors: a primary school and a Government hostel, for example, by Gerard Brigden, chief architect of the Colombo team, appear excellent.

For many of those living there, housing standards are markedly higher than those to be found elsewhere in Pakistan. More recent residential layouts show great architectural improvement on earlier areas. The administrative sector can claim buildings of striking quality. The setting is as beautiful as when it was chosen by the Site Selection Committee in 1959—indeed, as the intensive landscaping programme develops, a good deal more so. For the work of Derek Lovejoy and Associates, on the pictorial evidence so far available, there can be nothing but praise.

Yet this chapter of Islamabad's history ends with a query. Will the city turn in upon itself, to reflect an enervating nostalgia for the forms of other days, perhaps appropriate to a regional centre but hardly to a national capital? Or will the city, as it grows, body forth that vision of a capital, vigorous and varied, which would forge an indigenous twentieth-century architecture and reflect the aspirations and ideals of a young nation in the modern world? In Pakistan's battle of the styles, the present protagonists appear to be the Parliament Building and the Secretariat. Possibly the completion of PFR's Secretariat will prove the touchstone.

9. Interior Design

There are areas of design, as we have seen, in which efficiency of performance must be the paramount consideration. These are susceptible to wholly logical analysis, and the solution of their problems can be aided, in many cases, by sophisticated methodology and computer techniques. There are other areas, no less important to the well-being of man, which are not susceptible to exact measurement. In these it is the designer's visual sensitivity—his flair for form, materials, colour, texture, light and decoration—which takes precedence in the design process. Such fields of practice, and they are no less specialized than any other, include textile and carpet design, furniture, the mural painter or sculptor, lighting, silversmithing, glass and ceramics, typography and graphic design.

The meeting point for some, or all, of these is interior design. The orchestration of their skills and products in the public interior calls for creative thinking on a certain scale, smaller than the urban scale, broader than that of single craft-based industries. In so far as the brief is usually looser, or less demanding, than in purely functional design, it allows, and often requires, greater aesthetic inventiveness. Its constituent elements are seen in close-up and demand particular felicities of detailing and trim. Interior design has in consequence emerged as a speciality on its own and there is an increasing tendency for architects or their clients to call in such specialists to handle the internal spaces of their buildings. Hotels, restaurants, university common-rooms, board rooms and other office interiors, airports and railway stations (and indeed the interiors of the planes and trains themselves), clubs, cinemas —these are only some of the public and semi-public spaces which modern society requires for its business and its play. Affluence and mobility have transformed them. Between the inter-war tea shop and the controlled exuberance of the 'speciality' restaurant lies a social revolution. Nowhere is this more evident than in establishments catering for the tourist trade. Tourism has come to rank highest in the table of international industries. Competition is intense. The traveller with money in his pocket demands the best and nothing but the best will attract him. Infinite care is lavished upon his surroundings so that they will soothe him, excite him, wrap him in luxury, minister to his bodily comforts.

A clear-cut if rather special example is to be seen in the great liner. A ship is a floating hotel,* but poses additional problems of its own. Some of these

* Nowadays. Sir Colin Anderson has reminded us that the cabin was once a space you

are technical, for its services are complex and the need to save every cubic foot of space requires exceptionally compact designing. Some are psychological, for the close proximity of large numbers of people in a confined space can easily lead to claustrophobia, their enforced idleness to boredom. Among British shipping lines the Palm Court tradition dies hard but, during the nineteen-fifties, Britain began to catch up with the fresher design standards established by the Scandinavian countries. In this the Orient Line set the pace, as indeed they had been doing for thirty years, notably with their *Oriana*. *Oriana* is no longer a new ship (by the mid-sixties she had carried over 100,000 passengers) but she still holds every record east of Suez and is generally regarded as one of the best-designed ships on the high seas.

Many people worked to make her so. The essential credit, however, belongs to her owners, whose policy decisions made it possible. Further: it was one of their number in particular who guided the board through those decisions—Sir Colin Anderson, for long a notable figure in the London world of art and design. In his own words:

> More than one generation of the family had been deeply interested in good design and . . . it was taken for granted that a new project involving the element of design would be submitted to the chosen member of the management for discussion, and finally for approval, before any final decision was taken. It worked only because we were lucky enough, as a family business, to produce a series of members of the management who did prove to possess a natural capacity for fulfilling this function. At one time the job of choosing designs had been fulfilled by a Committee of the Board; but that was wisely abandoned and the final solution was to entrust all decisions about design to one individual who had, to assist him, a management trainee.*

'We were lucky enough . . .'. On such strokes of luck in industry and commerce turn expenditures of millions of pounds. Rare is the management in which the claims of design are represented at board level. Lacking such advice, great undertakings can be, and are, set upon disaster courses from which no later salvage operation can save them. Immense and hideous buildings are commissioned by well-meaning men who know not what they do. Hundreds of square miles of urban Britain have been degraded by items of street furniture that are brutally ugly—produced by companies and installed by local authorities who have seen no need to take adequate advice. In the very world we are here considering, the world of shipping, the rational

furnished to your taste. On the *William Metcalfe*, sailing to Australia in 1839, 'passengers were recommended to have their cabin furniture made and fixed before starting the voyage. A Venetian door of jalousies was also strongly advised . . . there can have been very little privacy and precious little comfort here'. Thomas Gray Memorial Lecture at the Royal Society of Arts, 2 March 1966.

* Sir Colin Anderson, speaking at a COID Design Congress in 1961.

co-ordination of the interiors of the Cunard 'Q4', a ship likely to cost no less than £30 million, was secured only by board changes at well-nigh the twelfth hour.

If Sir Colin Anderson and his fellow directors constituted, so to speak, the joint producers of the *Oriana* design story, its joint directors were Design Research Unit, the team appointed to co-ordinate the design effort. Before following the story through in detail, it may be of interest to glance at DRU's background and the attitudes they brought to their task. A convenient starting point is provided by an office discussion meeting.

It is 10.30 in the morning. There are 38 people in the room, three deep round the tables with a couple of latecomers stacked near the door. At the 'top' table sit the senior partners. Against a cork pin-up wall at the opposite end an architect explains the reasoning behind two long elevational surveys of the buildings in Gold Street, Northampton. These are to be the subject of a collective face-lift on the lines of that first sponsored by the Civic Trust in Magdalen Street, Norwich. Photographs of the street are being passed round from hand to hand as he talks.

'There are two points I think we might usefully discuss', says Misha Black from the chair. 'First, is it justifiable to expend so much energy on an operation of this kind in a street of this kind? Should one just leave it to stew in its own juice until the time comes for its proper redevelopment? Second, if not, is this the kind of treatment a street of this character calls for?'

Someone says they feel it is a valuable exercise that can do nothing but good. Someone else expresses misgivings about a particular lemon yellow in the colour scheme. 'It will all have toned down tremendously in a year's time', says a third. 'Surely it ought to be *much* more unified', says the second voice; 'I don't feel the scheme pulls the buildings together.' 'But every building is different', says the architect; 'you really must accept diversity in a street of this kind. That's the sort of street it is.' There is a brief discussion about the readability of vertical signs outside the shops ('Chinese restaurants, I presume', says a senior partner) and the opinion is expressed that readability scarcely comes into it: such signs are read as symbols or logotypes, rather than letter by letter, and indeed they might well be considered to a far greater degree as decorative abstractions. On an inconclusive note discussion moves to the next item on the agenda.

This entails consideration of new uniforms designed for British Rail and already the subject of press publicity. Four members of the meeting slip out to act as models. While they are changing, a senior partner-in-charge sketches in the background. 'It was', he says, 'an open brief. A principal aim had been to produce an easy-fitting uniform for all ages, sizes and shapes, functionally efficient, based on current casual clothing yet sufficiently smart to be accepted as a uniform and to remain so however worn, in order to improve the morale of station staff and attract a better grade of recruit.' A design team of three, with an administrative assistant, had formed a working party with representatives of British Rail. They had travelled in nine European countries to study railway uniforms elsewhere and had received information from the United States and Canada. British Rail were undertaking tests on the wearing qualities of different cloths.

The 'models' return. They are wearing the proposed uniforms for a porter, a collector, an inspector and a station master. 'I suggest', says the partner-in-charge, 'that these should be discussed as far as possible through the eyes of the public, rather than on technical design grounds.' One of the design team introduces them. 'It seems', he says (referring to press comment), 'that to the typical Englishman these appear a typical Continental style. . . . In actual fact there is no resemblance *at all.* We have pictures here of every Continental uniform' (they are circulated) 'and you will see that the German uniforms, for example, to which the newspapers compared these, are totally different.'

The Chairman interposes to say that he feels the caps, or hats, in particular *do* have Continental associations. 'We *could* have kept more closely in line with British military uniforms—in fact the first caps *were* in the British railway, police and fire service tradition, but the British Rail Design Panel wanted a complete change and asked for them to be reconsidered.' He also suggests that there is an element of musical comedy in the braiding. 'Need there be so much gold?', someone asks. The designer says there had been earlier thoughts of regional differentiation here but these had had to be dropped. He points out that improvement of morale is one of the purposes of the proposed changes. 'In the French navy', says someone, 'the higher you go, the less braid you get. At the top you get none at all.'

Somebody thinks the station master is the most successful. Someone else admires the porter's cap as being the most original. 'Wehrmacht', says another. 'I agree that the porter's cap is the most successful', says the partner-in-charge, but in fact British Rail have asked for this to be brought into line with the others.' Proposed modifications in the porter's uniform, eliminating the gold and restricting it entirely to charcoal and red, are described. Questions are asked about the buttons ('do they have to be polished?') and the wearing quality of the materials (British Rail's tests showed wool to be better than a mixture of wool and synthetics) but feelings are clearly centred on the caps.

It is the *height* of the collector's and the inspector's caps, someone says, that leads to German connotations. 'They are all cut the same', replies one of the design team, but the evidence of the eye seems to refute this. The Chairman sums up. 'I have all along felt', he says, 'that the badges for collector and inspector' (they are round) 'are too big for the caps, with the result that they force up the front and give rise to this Continental association. The station master's, on the other hand, with its straight cut, has a nice, rather Victorian feel (the station master's badge is a horizontal ellipse). If the braiding cannot be changed' (it had already been approved once by Dr, now Lord, Beeching, the I.C.I. director borrowed by the Minister of Transport to reshape the outworn British railway system), 'then at least you might give the curve a tighter radius. I think one or two small changes like these will improve the uniforms out of all recognition.'

Discussion now moves on to a series of eight surface designs for printing by litho on to shallow pressed metal trays intended for the mass market. These represent an attempt to set a new, and relatively sophisticated, standard of design in the very lowest price range ('Fornasetti at cut price', says someone). The designs include a big Victorian rose, an abstract of overlapping circles done in various colourways and a rather jolly pussy-cat. The rose, it seems, took the place of an existing design which was suppressed by the manufacturer at DRU's instigation and consequently never marketed; the others will be submitted to the manufacturer as part of a general plan of suggested developments. 'Trays are for use', says Sir Herbert Read; 'these are too jazzy to have a cup of tea on them.' Another partner says: 'I think the rose is awful, inexcusable.' 'You ought to change the tray first', says someone else, 'it looks so cheap.' The designer explains that the rose was by way of being an emergency saving move, a stop-gap, and that the manufacturer had transposed the colours. He says too that they are changing the rim of the tray with a PVC extrusion but that it will take some time for the manufacturer to change his machinery. 'Who did the cat?', someone asks. The designer says that he first did a cat design himself but, feeling it to be unsatisfactory, had got this done by someone outside the office. The feeling of the meeting is clearly that further thought should be given to both the form of, and the surface designs for, these trays but that if the general aim can be realized, it will indeed mark something of a breakthrough in this particular market.

Three more subjects are on the agenda: A building problem in Boston, Lincolnshire; a range of packaging; and the entrance hall to some new offices. Time permits consideration only of the first. The property concerned, it is explained, consists of an agglomeration of buildings and outhouses with an enormous hinterland behind the present frontage. The owners would like to pull the lot down; the local preservation society are opposing this. DRU have been asked by the owners to advise on the relative merits of conversion and redevelopment, and on the sort of building the latter might involve. It is felt in fact that a completely new building is the only possible solution.

Discussion is fairly intense about the scale of the proposed building, its relationship with its immediate neighbours and its somewhat negative character.

The partner-in-charge explains that they have already had to persuade the client to accept a two-storey building instead of the single-storey development preferred; and that the character of the job (brick, stone and pitched roof) is quite deliberate. 'We gave a lot of thought to whether a spanking modern building would be right here. We came to the conclusion that on this postage-stamp site the town simply couldn't take it.' There is obviously unease here and there that the building as so far envisaged will fail to make a real contribution to the town, but, equally, a feeling that the general approach is nonetheless correct.

The time is 12.30. The meeting has lasted two hours. An announcement on new work accepted by the office is made (an exhibition at Lima, in Peru; a house style; stationery; an electricity substation) and everyone disperses for lunch.

The six-weekly design discussion meetings of Design Research Unit, London, of which this was a fairly typical example, are central to its philosophy and reflect an idealism at once amiable and shrewd. DRU is the largest practice of its kind in Britain and handles all types of design from the purely architectural to the purely graphic. For Gilbey's, the wine merchants, it first designed bottles, labels and packs; was asked to handle the interior design of their retail shops; and ended up advising on the architecture of a new distillery, stores and an award-winning head office. For British Railways it has designed locomotives and rolling stock and devised a new name—'British Rail'—a new symbol, new liveries for coaches and ships, a new house style for print and publicity. Other clients have included the GPO (a new look for post offices), London Transport (advice on the new Victoria Underground line), De La Rue (advising on the design set-up for a whole group of companies), Ilford Ltd. (a new house style), BP (office interiors) and the Central Office of Information (overseas exhibitions). The Unit is neither a firm, nor a partnership, nor yet a loose association of individuals, but contrives to embrace elements from all these.

As earlier indicated its origins go back to the war years and beyond. Three present partners—Misha Black, Milner Gray and Kenneth Bayes—were among those who formed, in 1933, a group called Industrial Design Partnership. With the coming of war, this inevitably ceased operations. In October 1942 Milner Gray, by then working, like Black, for the Ministry of Information, prepared a note outlining the framework of a service 'so complete that it would undertake any design case which might confront the State, Municipal Authorities, Industry or Commerce'. Three Divisions were envisaged: Design and Fabrication, Technics and Research, Finance and Administration. Associated with a whole-time nucleus of experts was foreseen a wider circle of specialists on retainer. This paper found support with a wartime federation of smaller advertising agencies, the Advertising Service Guild, which agreed to sponsor an embryonic office capable of expansion when peace came.* The Guild member primarily responsible, Marcus Brumwell, remains a partner in DRU to this day. Thus was born, in the spring of 1943, Design Research Unit. Its part-time Director, a little unexpectedly, turned out to be Herbert (now Sir Herbert) Read, poet and art historian, whose *Art and Industry* had earlier evidenced his interest in industrial design. To guide him he had the advisory panel of specialists brought forward from Gray's paper. To assist, at a slightly later stage, he had the retrospectively even more surprising figure of Bernard Hollowood, now editor of *Punch*.

The Unit's objects were set out in a number of leaflets (written by Read). They are worth quoting simply because the aspirations they embody were rare to the point of non-existence, at least in Europe, at that time. Among its aims were: 'To carry out research into the needs of the consumer and the potentialities of industry . . .; to secure the co-operation of artists and designers in projects involving more than one material or more than one purpose; to bring such artists and designers into productive relation with scientists and technologists; to collect and correlate information about industrial design from all sources; . . . to create a contemporary school of design.'

In the cold light of the nineteen-sixties, this sounds like an outline programme for the Hochschule für Gestaltung at Ulm, the Council of Industrial Design and the International Council of Societies for Industrial Design combined. It must be remembered that in 1943 none of these existed, even in imagination. What is interesting about the manifesto is its groping towards ergonomics and technical research, its recognition of the designer as only one member of a highly qualified *team*, and its aim to build up the Unit on a co-operative basis. Implicit, too, is the social sense which DRU has contrived in some measure to carry down the years.

'It is necessary', said that first four-page leaflet, printed at a time of rationing, blackout and general austerity, 'to re-integrate the worlds of art and industry, for only on that basis can we progress towards a new and vital civilization.' 'We shall not be satisfied until we have not only an office and a coherent group of designers, but also a laboratory-workshop where designs of an experimental nature can be evolved. We wish to lead industry in the field of design and not merely follow where a demand has already been created.'

During 1943 and 1944 the Unit began to get under way and a number of jobs were undertaken. A pointer to its general approach may be seen in the car design—for Alvis—which Herbert Read persuaded Naum Gabo, then resident in Britain, to undertake. The design proved unacceptable and was never put into production but is said, as one of the first hard-topped, two-tone designs, to have been 'ahead of its time'.

* The Guild was also responsible for bringing 'Mass Observation' into the field of consumer research.

Early in 1945 first Milner Gray and then Misha Black were released from the Ministry of Information and became full-time members of the Unit, to be joined a year later, by Kenneth Bayes and Dorothy Goslett, an administrator with whom all three had been closely associated at the MOI. The outer ring of advisory panel members were rechristened, more pompously, the Design Directorate and now embraced, besides Black and Gray, Norbert Dutton (designer), Frederick (now Sir Frederick) Gibberd (later architect-planner for Harlow New town), Felix J. Samuely (engineer, and notable for his work on architectural structure) and Sadie Speight (architect and wife of Professor Sir Leslie Martin, already mentioned as Architect to the London County Council).

A brisker note of realism now became apparent in the Unit's publicity material. 'Successful design', states an October 1945 leaflet, 'is based on data . . .'. Reference is made to materials and manufacturing processes, to the cost and efficiency of the former and the economic possibilities and limitations of the latter; to markets, competition and new fields. 'This research', states the leaflet, 'is part of the design process.'

The same leaflet reiterates the principle of group working. 'Like every aspect of modern industry, design should be a co-operative activity and the function of DRU is to focus on every project it undertakes the combined knowledge and experience of several creative minds, since it believes that only by pooling the talents of a team of designers is it possible to offer a service capable of meeting every demand from the wide and varied field of present-day industry.'

And so, in fact, did it work out in practice. All inquiries came before the Design Directorate for acceptance and allocation to designers; save in cases of particular urgency, no solutions were submitted to the client without the Design Directorate studying the problem and tabulating criticisms. A quorum of the Directorate consisted of not less than five members; decisions were by a simple majority. As the nation felt its way back to peacetime conditions, however, the architect and engineer members of the Directorate found their private practices increasing to an extent which made it difficult for them to accept DRU assignments, then even to attend meetings regularly. Conversely, the 'inside' members found their work developing to the point where increased staff and space were necessary.

In 1947, the Unit moved to larger offices. The whole-time staff increased to twenty. The original, loosely-knit group of inside and outside members—those of them who wished to remain with the Unit—consolidated into a professional group practice of architects and designers, with an outer fringe of associates, some, but not all, of whom actually had their offices within DRU. These latter were called upon as and when appropriate to collaborate in, or be completely responsible for, jobs for which their experience made them particularly suited.

six months' time the keel of a new liner, *Oriana*, will laid down. Above, a wooden model, developed from rlier versions in wax, is under test in the shipbuilders' perimental tank for evaluation of her hull form. The del is self-propelled; the travelling rig above carries the instruments for obtaining performance figures. Wind tests are taking place concurrently at the National Physical Laboratory. Below, the way the Ship Drawing Office saw the proposed vessel at this stage.

3 November 1959: *Oriana* is christened and launched by HRH Princess Alexandra at Barrow-in-Furness. Another year remains before her maiden voyage. No fitting-out of her public rooms has started, but design work on her interior has been in progress for two years. There are numerous innovations in her layout—for example the stepped plan, right, of her court cabins enables even inside cabins to receive natural light and a glimpse of the sea through their corner windows on to the court (plan, right, shows part of 'C' Deck. *Oriana*'s sides are the top and bottom edges of the plan).

Over the next twelve months 100,000 sq. ft. of public rooms took shape under the co-ordination of Design Research Unit. Opposite above, the First Class Restaurant, by R. D. Russell and Partners. The floor—garnet and black linoleum; the chairs—red hide; the carved and gilded chandelier is by Professor R. Y. Goodden; the relief at the far end by Geoffrey Clarke.

ˈlow, the Tourist Ballroom by Ward and Austin. ıe general colour scheme is orange and green; corative panels, on the theme of 'The Masque of iana', are by Olive Sullivan. Who would guess that a main 'trunkway' passes through the middle of this room and that, in port, large sections of floor and ceiling swing away to enable bulky cargo to enter and leave the hold several decks below?

Above, the Junior Club—one of two children's areas, each allowing for quiet and noisy activities. The climbing frame (seen in close-up below right) was designed by June Lyon of DRU and tested in a children's club in Shoreditch. Decorative plastic panels (see also opposite, top left) by Robert Perritt, also of DRU. Below left, a detail from the cinema, showing the circular staircase between balcony and stalls.

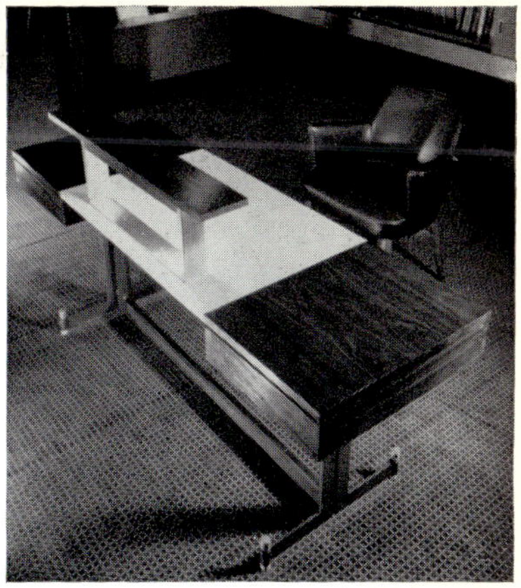

ome other details. Top centre, John Piper's long ıural, screening the First Class lounge from its brary, and, right, the librarian's table. Above left, orner of a Tourist lounge; right, relief by Mary Martin on the main stairway. Similar reliefs on each landing identify the deck. Below, one of the illuminated Deck Plans, by Christopher Timings of DRU—acid-etched on the reverse of glass, colour filled and fired.

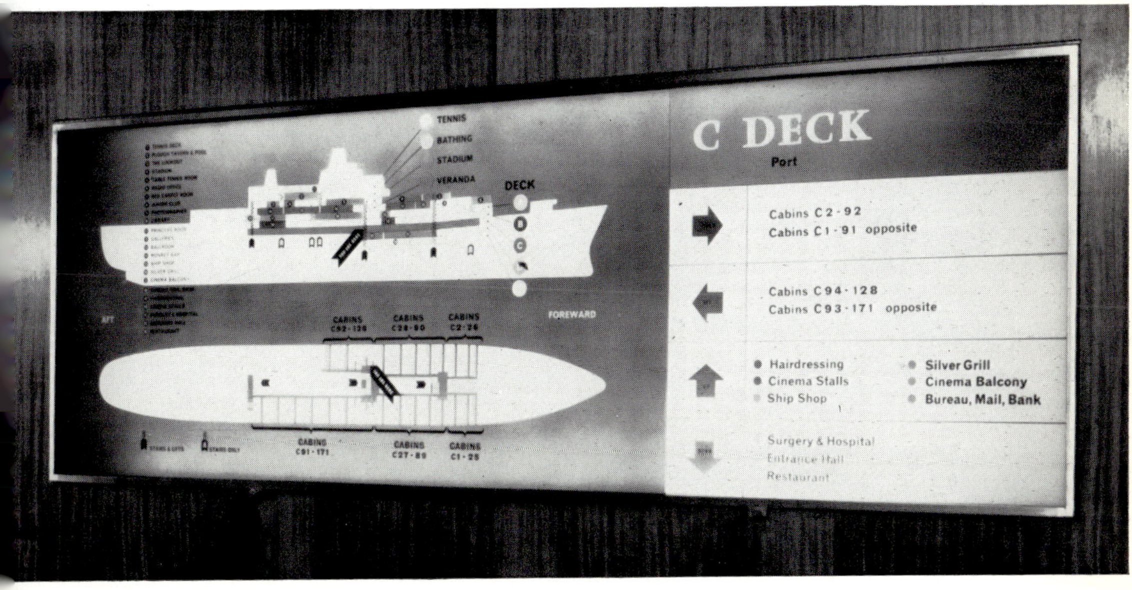

Above, showcases at the entrance to the Ship Shop. Beyond, another Deck Plan. Colour coding and arrows inlaid in floors were part of a co-ordinated system to help the landlubber identify his course and position on any one of eleven decks. Partner-in-charge, Milner Gray. Below: plaque at the First Class embarkation entrance, compass in the First Class lookout, and waste-paper bins.

ɔove, the Stadium Lookout, with its fine views aft, s the most nautical feel of all the public rooms. ılkheads are in cherrywood veneer and blue plastic ced panels bearing the ship's crest; floor dark ey. Right, the First Class bathing pool; mosaic by bert Stewart.

Following page: a view of the upper decks and dge wing. The use of light alloy in the superucture and the placing of the lifeboats, traditionally rried on the top deck, some 30 ft. lower down, so proved the stability of the ship that an extra deck uld be introduced.

Lower picture: at Palma, *Oriana*'s lines show more arly and may be compared with the painting on ge 205. Apart from the inflation of scale usual to ch pictures, derrick and Samson posts have gone d considerable modification to the superstructure be apparent. Six years after her maiden voyage, *ana* still held all records east of Suez.

ORIANA

During the next ten years, staff and associates increased to about 40. By January 1966, the full-time staff consisted of 3 founder partners, 9 other partners, 19 architects and interior designers, 13 graphic and industrial designers, a librarian, a production assistant, with 18 administrative and general office staff. Three associates remained closely linked with the Unit, although only one retained an office within DRU. In addition the Unit consistently calls back for collaboration some dozen or so ex-members of the staff who have at different times left to go into private practice. From 1954 to 1959 DRU had an office in Dublin; a small office is now maintained in Newcastle upon Tyne.

Over the years, Design Research Unit has developed from an organization of young rebels to become a pillar of the establishment. Some would say that the youthful fire has gone out of it and that the quality of its design—efficient, tasteful, carefully detailed, thoughtful and thorough—smacks too much of the establishment too. No one could accuse DRU of a house style—its designers are too conscientious to impose preconceived notions upon the jobs they undertake—but it is perhaps true that in all their varied output surprises are nowadays rare. There are no Gabo cars, nor likely to be. Failure to startle is, however, a complaint more readily levelled by designers than by clients or public. In the case of DRU, certain factors have perhaps operated with particular force to create a marked group mentality, a common approach to the problems of design.

In the first place, there has been an exceptional continuity among the senior members. Three partners, as we have seen, have been in practice together for well over thirty years; two more have been continuously associated with them for over twenty; other partners and associates have been linked with the Unit for fifteen years or more. This alone creates a common datum of shared experience that is still rare in the field of industrial design. Nor is the pooled knowledge represented by long collaboration with others outside the office to be discounted.

Most practices, having accepted a particular type of job, attempt to recruit appropriate skills on to the staff; such assistants are largely unknown quantities. Even though DRU's 'associates' conception has largely withered away, most of the outside specialists with whom the office works have collaborated at intervals over a long period and both sides are therefore known and trusted. Those (a dozen or so) who were once on the Unit's staff speak the same design and administrative language; others enjoy the freedom of design approach and the degree of administrative help they receive.

Further coherence is lent to the Unit's corporate thinking by the Design Discussion Meetings, usually with Sir Herbert Read as guest critic, one of which has been described. Attendance at these meetings is compulsory for all creative and executant design staff, with representation by rota of administrative and clerical staff. The direct value of such meetings is variable; they may swing from the electric excitements of a 'brain-storming' session to the merely aimless expression of personal opinion. It is probably rarely that they provide any direct assistance for the designers concerned. Designers are in any case not required to adopt the suggestions put forward unless they wish to do so. On the other hand it sometimes appears that those most prickly in public defence of their projects are most prone subsequently to amend them in the light of the discussion. DRU's Design Discussion Meetings should perhaps be considered an error-correcting device of intermittent value but an undoubted instrument of education for its younger staff and, to a degree probably unrecognized by many of those concerned, a means of consolidating office morale and unity.

Staff is chosen for general ability plus some specialist skill. For graphic and industrial design work, the Unit employs a handful of assistants and juniors to detail for the appropriate manufacturing or printing process the design concept established by senior designers (though such juniors are generally art school trained and if they show creative ability are always given the opportunity to put this to use). The seniors themselves tend to a measure of specialization through personal sympathy for a particular material (for example, glass, plastics or metal) or by reason of a particular skill or cast of mind (leading to, say, packaging or typography).

In its architectural and interior design work, on the other hand, the Unit employs no draughtsmen-assistants who are not capable, under creative supervision, themselves of working out details practically and efficiently: and, if possible, with the experience and personality themselves to deal with clients, contractors and suppliers. On long-term jobs, teams remain stable over considerable periods but assistants do not necessarily continue in the same team subsequently, and there is no insistence on specialization where assistants prefer it otherwise.

Elementary research on ergonomics, materials and methods is undertaken by the design team as part of the job and as a matter of course. Specialist assistance is sought where specialized research is called for. The most junior members of any team are free to express a forthright difference of opinion but final authority lies with the senior in charge of the job. 'If the machinery is to function', says Misha Black with some regret, 'there is an inescapable necessity for someone to take the final decision. If I take full charge of a job and one of the team turns in something I don't like, it has to be changed—in theory. In practice, the thing is affected by all sorts of intangibles resulting from the group tradition. If the designers concerned feel strongly, I would probably say: "Let's try it out on the client and see what he thinks. If he doesn't like it, we'll try it again in such and such a way."' As far as one can judge, pitched battles are rare. As another member of the Unit put it, 'the people in DRU are intellectually' (it might have been more accurate to say temperamentally) 'related.'

The same stability is sought in the client relationship. Not quite to the same degree as Henry Dreyfuss in the United States but much more than used to be common in Britain, DRU works on a retainer basis. In such cases the Unit is able totally to identify itself, not merely with particular design problems as they arise but with the development of overall design policy over a period of years. An example that comes to mind is the exterior decoration programme established for the public houses controlled by the Watney-Mann Group of Companies, which hit the public and the trade at the time with the force of a revelation, and on which Milner Gray has been retained since 1956.

An obvious factor in the readiness of firms to commit themselves to the Unit for long periods and complicated undertakings is DRU's reputation for administrative efficiency in overall co-ordination. Its readiness to take such overall control in its stride is reflected in the comparatively high ratio of non-creative to creative staff. The readiness of other architectural and design offices on their side to submit to DRU's co-ordination can perhaps be ascribed to its tradition of working amicably with 'outsiders' and to the knowledge that, once a general design approach has been mutually agreed, they will have full design freedom in those parts of the job for which they are responsible. The way *Oriana* was designed provides an example of how this works in practice.

In September 1956 the Board of the Orient Steam Navigation Company (since amalgamated with its parent associate, the Peninsular and Oriental Steam Navigation Company, to form P & O Orient Lines) formally took the decision to lay down a new ship of 40,000 tons, for passenger service to Australia and the Far East and for cruising in off-season periods. This was ship No. 1061, later to be known as *Oriana*. By the moment of decision, the design process was already far advanced and many of the important things about *Oriana* were already known. The developing interaction between technical and economic feasibility studies and the known general requirements of the owners meant, indeed, that no clear-cut brief as such was ever drawn up. Rather were those concerned with the creation of this ship faced by a long succession of diminishing areas of decision, from the moment her immediate predecessor *Orsova* left Barrow to be fitted out in 1954 to *Oriana*'s maiden voyage on 3 December 1960.

In this long chain of decisions, the need for the first was stated briefly by Sir William Currie, then Chairman of P & O, who were concurrently facing a similar problem. The choice, he said, had been one of 'building the most economical and competitive replacements for our older ships or beginning to retire from the Australian passenger trade altogether'. In the changed, and changing, conditions brought about by air transport and new patterns of political power, that is to say, the design starting point was not to be found in an analysable demand but in a general requirement for greater efficiency and increased performance.

For long, the Orient Line had operated the Australian run. In 1954 the Company introduced a Pacific service and in 1955 two round-the-world voyages were made. These voyages nearly doubled the period away from London, from some ten weeks to about four months. First thoughts, therefore, were turned to the possibility of a faster ship. To chop a day off the long passages of the Australian-Pacific run—say from London to Port Said, or Suez to Colombo—speed would have to be increased from the 22½ knots of the Company's other post-war vessels—*Orcades, Oronsay* and *Orsova*—to 27 knots. At the same time it was clear that a ship of their size—28,000 tons—could not pack the power necessary to be driven at such a speed.

Discussion on these matters took place under the Chairman of the Orient Line, Sir Austin Anderson, whose special design team was composed of two Directors, Sir Colin Anderson and Mr Ford Geddes, the Company's naval architect Mr Charles Morris and, where appropriate, the Superintendent Engineer, Mr H. Knight. Preliminary calculations were complicated by the introduction of two possible design innovations. The first was whether to use an unprecedented volume of a light alloy instead of steel for the ship's superstructures, thus enabling a whole extra deck to be accommodated. The second was whether to gain certain internal planning advantages by siting the main engines aft—which in turn would enable the funnel, with its attendant smut and fume problems, to be sited aft as well.

The basic lines of various possible ships had been worked out by Morris. Final calculations could only be made by shipbuilders, however, and early in 1955 the Company therefore put the problem to Messrs Vickers-Armstrongs (Shipbuilders) of Barrow-in-Furness, who had built all Orient Line's then existing ships and were thus accustomed to working closely with the Company (and with whom, incidentally, Charles Morris had served his apprenticeship before joining Orient in 1944). What, they were asked, was the smallest vessel which could provide the performance envisaged?

After due study, Vickers-Armstrongs came back with proposals for a ship of 804 ft. overall, with a moulded breadth of 97 ft., driven by a shaft horsepower of 80,000 and of some 40,000 tons gross. In a closely argued set of recommendations, they discarded the design for the engines aft (it would have meant fewer passengers) but did advise the use of a very large quantity of aluminium. It was also suggested that the ship's lifeboats should be placed 30 ft. down from the top deck where they are traditionally carried. The effect of these moves would be so to increase the ship's reserve of stability that an extra deck could indeed be introduced at the top, with space for 150 more first class passengers.

Armed with this information, the owners now made some preliminary studies of possible passenger and crew numbers, capital and running costs. It became evident that:

(*a*) four such ships of 40,000 tons would produce the same number of passenger berths per year as six ships of 28,000 tons (i.e. the *Orcades, Oronsay, Orsova* class);

(*b*) four such ships of 40,000 tons would require £8 million less total capital investment to build than six of 28,000 tons;

(*c*) four such ships would require 500 less crew than six ships of 28,000 tons;

(*d*) the saving of time by the faster 40,000-ton ship would more than outweigh the cost of providing the extra turn of speed (this being mostly the cost of additional fuel).

These figures were decisive. In September 1955 an 'order' was placed with Vickers-Armstrongs. Completion was scheduled for 1960. Detailed planning now began.

This was the general situation when Sir Colin Anderson met Milner Gray by chance on Hampstead tube station around the turn of the year. Sir Colin mentioned that a new ship was being planned and asked whether DRU would be interested in acting as design co-ordinators.

As has already been made clear, Orient could claim a long-standing concern with design. As far back as 1912, the line had commissioned a well-known architect of the day to design interiors for their ships: the first such step, it is believed, on record. In the nineteen-thirties, many who had never been to sea nonetheless became aware of Orient's distinguished publicity, notably from the hand of E. McKnight Kauffer, for what was really Britain's first modern liner, *Orion*. It was in 1932 that the Company decided to employ a young architect, Brian O'Rorke, to design the interior of *Orion*. Thereafter from 1932 until 1954 O'Rorke acted for them in the internal (and to some extent the external) design of five ships. As the size of the ships increased, however, it became increasingly apparent that their complexities demanded a larger design team (Liners have been known to begin maiden voyages with more carpenters aboard than passengers). A new approach was necessary. For their next ship the company decided to employ 'some organization specializing in the co-ordination of large and perhaps unusual building projects' on the understanding that O'Rorke should still be responsible for certain important rooms and should act as overall adviser. Though DRU had had no previous experience of ship design, Sir Colin would undoubtedly have known of their reputation for efficient co-ordination in other spheres.

So it was that Misha Black met Sir Colin on 27 February 1956. From this meeting he brought away the client's brief: a 5½-page document titled *The Problem*. It is worth quoting that key document at some length. It already envisaged group working and listed the anticipated duties of the co-ordinating organization.

'The problem' was stated succinctly in the first sentence:

To get the interior of a 40,000-ton passenger liner designed and completed at, or before, the completion of the hull and engines. The document gave the anticipated launching and completion dates. By the launching date, rooms on the lower decks should have their steelwork completed. The upper rooms will have taken shape but will probably not be quite ready for decorative contractors to lift detail beyond that given in the structural plans. All structural plans should be available progressively up the ship and be complete before launching. Very approximate floor area is of the order of 70,000 sq. ft., involving possibly 26 public rooms of varying sizes, apart from stairs, landings, etc.

The co-ordinating organization would be charged with the basic task of seeing that the grand completion date is adhered to and that:

(*a*) Sub-contracts are placed in proper time;

(*b*) Sub-contractors are on the right path towards fulfilling their dates of delivery (in short, general progress-chasing);

(*c*) The necessary information is supplied to Builders' or Owners' sub-contractors concerned with such matters as ventilation, sprinklers, loudspeakers, etc., in proper time;

(*d*) The work of various designers is such as will compose into an harmonious total effect and have clients' (i.e. passengers') approval;

(*e*) Textiles, furniture, carpets, electrical and other fittings and decorative features and so on are being selected or designed in time and that they have the clients' approval.

It would be envisaged that the Managers of the Orient Line, in consultation as necessary with this co-ordinator, would appoint a limited number of architects to be responsible for various defined areas. Thereafter these architects would become the responsibility of the co-ordinator in so far as progressing the job was concerned and also in coming to an agreed basic design for submission to the Orient Line. It would also be understood that if any of this panel of architects should, for any reason, be unable to complete his functions, then the co-ordinator would undertake to do so, either directly or by introducing some approved substitute.

To particularize, the co-ordinator is to be the link between: *Group A* consisting of:

1. *The Management of the Orient Line* (Sir Colin Anderson, Mr Ford I. Geddes (assisted by Mr A. M. Stirling) and Mr D. J. Donaldson (Finance)) and their Naval Architect's Department (Mr C. F. Morris);
2. *The Builders* of the ship, with their various separate branches:
   (*a*) Ship Construction,
   (*b*) Electrical,

(c) Engineering,

and their sub-contractors;

and *Group B* composed of:

3. *Various Architects and Designers* responsible for particular sections of the whole. (The work is capable of infinite subdivision—though the more subdivision there is, the more difficult will be the job of co-ordination.)
4. *Sub-contracting Firms* who construct the interiors. (There are only a limited number of firms specializing in ship work.)
5. There also has to be arranged some orderly way of seeing to the proper choice and ordering of approved textiles, furniture, carpets and features, if for some reason these are not the responsibility of 3.

The document went on to refer to Brian O'Rorke's special knowledge of Orient Line requirements and stated that R. D. Russell was already working on cabin design.* Only public rooms and spaces were therefore involved in the programme envisaged. It continued:

The following shows, in very rough outline, a suggested method of dividing the work to be done on the ship into a series of categories, each one representing a certain kind of requirement and therefore, theoretically, capable of being dealt with separately from the rest. It is not suggested that it would be wise to farm them out between eight different architect-designers. But the list indicates that there are a number of alternative means of arranging the work into, say, three or four groups.

A point to be considered, though it is not a cardinal one, is the fact that ships tend to be completed from the bottom upwards so that it is easier for an architect and his decorative sub-contractor if their area of responsibility has something of a vertical arrangement about it, instead of comprising nothing but large projects on one horizontal plane and thus all likely to reach their culminating activity at the same moment.

The following categories were then listed in detail:

Category A	Cabins and similar spaces (first class, tourist and crew)
Category B	Circulation Areas (first class, tourist and crew)
Category C	Hospital
Category D	Cinema
Category E	Swimming Pools and Taverns (first class and tourist)
Category F	Public rooms (first class)
Category G	Public rooms (tourist)
Category H	Public rooms (crew)

After meetings with O'Rorke and various internal discussions, DRU wrote to Sir Colin Anderson three weeks later, commenting on the brief and proposing fees. Negotiations were concluded and confirmed in a letter of 3 May.

* He had in fact been doing so since August 1955—one whole year before the construction order was placed.

May–November 1956

The following months were occupied with clarification of the design and procedural issues involved. The first need was for the senior members of DRU concerned to familiarize themselves with the technical, functional and human factors that might affect the design problem ahead. Inspection was made of existing Orient Line ships, culminating in a week-long cruise by four partners (Misha Black, Milner Gray, Kenneth Bayes and Dorothy Goslett) in *Orsova*, the Company's most recent ship, in August.

Collateral discussions on the procedure to be adopted for co-ordination and administration culminated in a paper from DRU dated 16 October. The basic recommendations of this paper were followed in general though not in detail. It covered the ground under nine headings: formation of design team; date schedules and shipbuilders' drawings; meetings; budget control; estimating and contracting; liaison and supervision; variations to contracts; supervision of installation, deliveries and completion; intermediate payments and final accounts.

It was suggested that, after discussions with the proposed additional designers as to their availability, DRU should negotiate fees and exchange contract letters (the latter having been first agreed with Orient). A first skeleton date schedule was set out, with procedures for coding and circulating drawings. Means were suggested by which a global production budget figure might be reached on the basis of preliminary designs, which, with a contingency figure, should then become a firm budget. Alternative methods of tendering were put forward for consideration and DRU offered to act as a 'clearing house' for all designers in respect of the contractors, suppliers, artists and craftsmen from whom they wished to invite tenders and estimates. The particular purpose of this was to avoid the main contractors being overloaded beyond their capacity and small firms, or craftsmen, being approached by one designer when their specialized skill but limited capacity had already been 'reserved' as being particularly suitable for another area.

It was further agreed that consultation between DRU and Orient Line should be by means of a Steering Committee. This consisted basically of Sir Colin Anderson, Ford Geddes and Charles Morris for Orient Line; of Misha Black, Milner Gray and Kenneth Bayes for DRU; and of Brian O'Rorke as consultant, with other members of the design team attending as necessary. First meetings were quarterly and relatively informal. Geddes had strong views on the practicalities; Morris doubtless had a naval designer's understandable wariness of land designers on the frontier of his territory (he it was, after all, who had to satisfy the Board of Trade's requirements governing stability, freeboard, load line, fire protection, watertight subdivision, and so on; who had to meet the Lloyd's Register rules relating to the strength of the ship and the quality of the machinery).

Sir Colin Anderson, in effect, had the last word on design. 'This was one of the few cases,' say DRU, 'the one astonishing job, where one felt the client spoke the same language and there was no need to go carefully . . .'. The first minuted meeting of the Committee was held on 26 July 1957; the last on 6 September 1960. Over this period 45 meetings were held in all, at monthly intervals, leaving as a record 151 foolscap pages of minutes. During the final twelve months it became necessary to supplement the policy-making functions of the Steering Committee with a series of purely executive progress meetings with all the contractors involved in the interior of the ship. These meetings—thirteen in all—were held in Barrow, for the most part monthly but during the final stages fortnightly. As time went on, the numbers attending not unnaturally rose: in October 1959 there were 17 present; by February 1960 there were 24; in July of that year 27; by August, about two-and-a-half months before the ship was to sail from Barrow, the number had risen to 30.

November 1956–May 1957

The Committee's first task was to assemble the design team which would together handle the entire interior of the ship. Four main offices were lined up—DRU itself, in addition to its co-ordination duties; Brian O'Rorke, in addition to his general consultancy; R. D. Russell and Partners, in addition to their work on cabins (traditionally the responsibility of the shipbuilders and in this case outside the terms of DRU's co-ordination) and Messrs Ward and Austin. Various proposals were made for the allocation of different spaces to these designers; the work was finally divided in the following approximate proportions by area: DRU 51 per cent; Ward and Austin 27½ per cent; Brian O'Rorke 15 per cent; R. D. Russell and Partners 6½ per cent.

Meanwhile, elsewhere, two years' intensive work on the form and structure of the ship were beginning to bear fruit. In the final event, *Oriana's* vital statistics were precisely as Vickers-Armstrongs had first indicated them in mid-1955. Before the interior designers could begin to plan, however—indeed, before the keel could be laid down—much development work had to be done before those initial assumptions could be regarded as final and their implications given precise form. It was, in fact, not until June 1957 that model tests in Vickers-Armstrongs' experimental tank at St Albans were completed and final hydrostatics produced. The ship promised to be exceptionally stable and seaworthy. Concurrently, preliminary schemes for her structural design were proceeding. Particular thought was given to the elimination of vibration and a scheme was finally devised of longitudinal girders and transverse webs and pillars. Into this framework had to be fitted, by trial and error, optimum passenger and crew accommodation and other necessary spaces.

In the solution of this multiple equation, many factors had to be considered. Some were capable of precise calculation; others were less tangible, though none the less real. Among the former, for example, was the simple fact that every increase obtained in passenger accommodation had to be matched by a corresponding increase in accommodation for the additional crew required to service them. Among the latter, the fact that standards of living and comfort are all the time rising, so that passengers demand more space. But how much more space? And how much more are they prepared to pay for it?

Nor were such things confined to the passenger areas. Before the war it was not uncommon for crews to sleep in large communal spaces holding up to 32 berths; in all but a few cases, *Oriana*'s crew were to be allocated cabins (with hot and cold water) for one or two men only. The effect of this upon design may be gauged by comparison with a pre-war ship. When *Orion* was reconditioned after war service the area of covered deck for crew use was increased from 23,350 sq. ft. to 33,160 sq. ft.—an increase of 42 per cent. Applied to *Orion*, *Oriana* standards would have meant an increase of 53 per cent.

Among the many other factors which bore upon the design of *Oriana* were, at random, the following: the rate at which cargo is taken in or out of ships has fallen; it was therefore decided to install mechanized conveyors for luggage and cargo. The ship was to have stabilizers and she was to be the first Orient Line ship that was fully air-conditioned throughout.* Fire precautions and the need for reserve stability against side damage had to be allowed for. She was to be given a novel system of transverse propulsion, controlled from the bridge, to facilitate docking. Because of the nature of her trade, more deck space for games had to be found than on, say, a transatlantic liner, and difficulty was experienced in reaching a layout which produced the right areas of deck at the right places. Further, research had to be conducted in wind-tunnel tests at the National Physical Laboratory in an attempt to ensure that such areas would be free of smoke and soot from the funnel, and from high winds. From electrical measurements taken at the intersections of grid lines on the deck, detailed charts were built up showing the wind velocities that could be anticipated at every point. At the same time the shape of the bridge fronts and other superstructures were developed and tested by the same techniques.†

A passenger ship with a quick turn-round must not be delayed by loading

* Stabilizers permit the strength and weight of furniture and fittings to be reduced. Air-conditioning, Sir Colin Anderson has said, 'has done more than any one other improvement to alter the interior design of passenger vessels making hot-weather voyages'. Rooms used to open up to embrace the deck; now they are shut in and steal the deck space. Royal Society of Arts Thomas Gray Memorial Lecture, 1966.

† 'But the fact is that the NPL cannot simulate all the magnified vagaries of the winds when they get tangled up around a huge funnel and in the complex marine architecture of bridgehouses, stepped layers of decks and the great moving cliffs which are the hull itself. Our enclosed deck is still subject to the strangest aerodynamic phenomena.' Ibid.

and unloading. Estimates on the desirable reservation of space and weight for cargo were therefore based upon the time required for embarking and disembarking, crew leave and necessary maintenance. When this had been agreed, the remainder of the space was broken down into parts to be given over to passengers, to crew, to baggage, stores, fresh water and fuel. It was estimated that 650 tons of fresh water would be required by the ship each day; problems of space dictated that the bulk of this be distilled by the ship's machinery. Conversely, research into the economics of buying fuel oil at cheaper ports and carrying it until consumed, or buying it at whatever price as and when needed, showed that big savings would be made possible by the former course. Provision was made, therefore, for carrying a large volume, 7,000 tons, of fuel oil.

By the spring of 1957, the main lines of *Oriana* and the general disposition of her interior had sharpened into near-finality. She would have an unusual bulbous bow, which would help to reduce pitching. She would be propelled by twin screws. She would sport one funnel and an engine and ventilation exhaust vent. The mast, which disappeared in her predecessor, *Orsova*, would return to support twin radar scanners. Derrick and Samson posts would be replaced by light electric cranes. She would carry about 2,000 passengers and 900 officers and men. The allocation of deck space to their many and various needs—cinemas and swimming pools, hospitals, restaurants, galleries, laundry, engine rooms and the rest—was near complete. She would cost £12½ million, or half as much again as the whole of Orient Line's pre-war fleet. As yet aesthetics, as such, had not entered the problem.

The designers were, however, beginning to receive information upon which their first thoughts could be formulated. DRU now had two clearly differentiated functions: that of general co-ordination and that of directly designing a number of spaces in detail. By the spring of 1957 preliminary work on the latter had begun. Within DRU it could be said that two teams now became involved: one was concerned with the interior design of the rooms; the other with signposting and deck identification (of all areas, not merely those which were the direct responsibility of DRU). Final design responsibility for the former team was taken by Misha Black; for the latter by Milner Gray. Overall co-ordination of progress and administration in relation to working drawings, contractors and so on, was the responsibility of architect Kenneth Bayes. Costing and purchasing responsibility was undertaken by Dorothy Goslett.

The group working on the public rooms varied considerably during the period of the job. For most of the period, there were about six architects or interior designers involved. The general procedure was for each of the assistants to take basic responsibility for one or more of the public rooms, with or without help, and to see them through from the design stage to final completion on the ship. One significant new appointment was made specifically for the job: a designer with previous ship experience with Cunard. As the only member of the team with such experience, his knowledge of ships was used in respect of *all* spaces concerned. One space, the cinema, DRU asked an outside firm of architects to design in collaboration with them; it may be noted that two partners of the firm in question had previously worked inside DRU.

The graphic and identification group consisted of four designers, headed by a team leader. Both teams worked together at various points, notably on the deck identification scheme.

May–November 1957

During the summer of 1957, the gap between the specifications of those shaping the structure of the ship and the tentative assumptions of those responsible for her final appearance began to narrow. By July, prototype furniture for the tourist dining room, for example, was being inspected. The centre to centre spacing for each unit (i.e. between centres of partitions) was agreed; the length and width of the table was agreed (though the length, it was thought, could be reduced by 1 in. if required); the chair backs, it was felt, should be reduced in thickness at the maximum point; the seat width was agreed though the dimension from front to back might be reduced if necessary for knee clearance—and so on.

By 3 September the Steering Committee was considering window types, the advisability of keeping within the ship's structural framing, the need to watch Ministry of Transport's required sill heights, and the necessity for due notice to be given in regard to any large glass areas so that the portal frames could be incorporated into the structure. Among other matters agreed at the same time were that there would definitely be no piano in the first class library and that pattern staining, caused by rotary polishers, would be reduced by the use of smaller tiles in the floor finishes. The 'double E' symbol —of outward facing entwined E's (signifying the two Elizabethan reigns) surrounded by a large O (the name *Oriana* deriving from one attributed to Queen Elizabeth I)—was agreed, subject to approval from Buckingham Palace, but the exact placing of the bow name, in relation to hull colours and portholes, was to be left open until a model was provided by Orient Line and until trial and error chalking on the actual ship could be effected.

A fortnight later, on 18 September, at Barrow-in-Furness, ship No. 1061 was laid down.

The programming of the job was complex. The shipbuilders initiated a progress schedule which was finally agreed in consultation with DRU. This consisted of dates against each stage of development for each public room. The schedule stipulated almost continuous 'to and fro' submission of drawings between the shipbuilders and the designers at every detail stage of development in relation to the shipbuilding and design of services such as ventilation. At each of these stages, either the ship-builder or the interior designer added their comment and additional information.

Design policy was established through the machinery of the Steering Committee. This operated smoothly and proved an effective method by which the designers could first be briefed by the client and secondly obtain approval of the decorative scheme for each space, generally in two stages: first, the general scheme; then the details of colours, veneers, finishes, fabrics and furniture. DRU formulated the agenda for each of these meetings according to the spaces for which schemes had been prepared, working upwards from the lower decks in accordance with the progress of the work on the ship.

Much research was now in hand; research into the real requirements of different age groups for the children's playrooms; into fireproofing of fabrics; into optimum light intensities for different purposes (readings from *Orsova* were obtained for comparison); into the artists likely to be used by Sir Hugh Casson in the P & O ship concurrently under construction in Belfast, so as to avoid any overlapping of commissions. Much thought was given to reducing weight by the use of synthetic and untraditional materials as deck coverings and non-structural fittings. But now already Barrow was tabling its demands and design was reaching its first points of no return. In October the shipbuilders were calling for the outline drawings and positioning of staircases, so that in turn the steelwork drawings could be prepared.

November 1957–November 1958

Throughout the winter and the following spring the many processes of adjustment continued. The designers adapted their thinking to the problems posed by long, low spaces up to 100 ft. square yet only 7 ft. 6 in. high. They learnt about camber and sheer, that the central area of a modern ship is pretty rectangular and that to expose the structure of the ship could mean no more than revealing a visceral tangle of ducts and cables. It was borne in on them that engine-room and cargo trunkways, running vertically through the ship, tend to impose upon large rooms an H or U shape. Revised passenger estimates by the owners led to modifications of accommodation—for example, an increase in the seating capacity of the tourist dining room by Ward and Austin from 1,520 to 1,544, to be handled in two sittings, or 12 more seats. By the same token the number of places in R. D. Russell's plans for the first class dining room could be reduced from 364 to 327. As it transpired, however, the revised plan would not work because of the disposition of crew quarters which could not be moved, so that further modifications were necessary, with further chain reactions.

Thus problems were shuttled backwards and forwards between designers, shipbuilders and owners, until resolved by mutual compromise. Here is a real difficulty posed by the positioning of some sandpits for children—three decks down from where the sand will have to be on-loaded; normally the sand would be transferred by chute but such is the position proposed that this will not be possible. Here, two portholes will have to be omitted to allow space for necessary service pipes. There, a colour scheme is considered too sophisticated and sombre for the earlier part of the day and the designers are requested to think again. What depth of water is permissible in the swimming pools, if the walls are of such-and-such height and a 3° roll is considered possible even with stabilizers?

By September 1958 DRU have been asked to draw up a list of decorating contractors. In Barrow, steelwork is 71 per cent erected or in production; half of 'E' Deck has been erected, 11 per cent more of the vessel's steelwork is ready, 29 per cent is in the pipeline; aluminium will not start production until February. A year remains before No. 1061 is launched; two years before her maiden voyage.

November 1958–November 1959

At the beginning of December it was agreed in principle to ask three contractors to work on the major areas of the ship. DRU was to request them to reserve capacity and negotiate terms. A firm of quantity surveyors would check the accounts. By the new year it was already becoming apparent that schedules were going to be very tight: in some cases only eleven months would be available between the completion of final steelwork drawings in Barrow and No. 1061's maiden voyage. In February the Steering Committee arranged its first visit to Barrow, in order to discuss and set up appropriate progressing machinery for the physical work on the ship's interior. In April 1959 the contractors were appointed. They were four: Messrs. Waring and Gillow, H. H. Martyn and Co., Bath Cabinet Makers and A. Edmonds and Co. Like the designers, they were allocated vertical blocks within the ship to ensure continuity and an even spread of work.

The contractors had previous experience of ship work and this greatly assisted the designers. Occasionally spaces were designed in right angles and it was left to the contractors to adjust these for sheer and camber in the light of the shipbuilder's constructional drawings. In certain technicalities of insulation, fireproofing, bulkhead doors, services and so on, there was dependence upon the knowledge of the shipbuilders and contractors: if they said a thing would work, the designers had confidence that it would.

In October the first progress meeting of the decorating contractors was held at Barrow. Present were 17 representatives of Orient Line, Vickers-Armstrongs, Ward and Austin, the three main contractors and Thermotank (who were responsible for that biggest of problems, ventilation); in the chair was Kenneth Bayes of DRU. Discussion here turned about schedules, dates, taking off sizes from the structure, setting out, responsibility for dimensioned drawings for lining off on the ship, commencing grounding and all the mechanics by which the designers' conceptions would be clothed in reality.

At the same time, in London, the Steering Committee had long passed this point in their deliberations. They were, for example, deciding that of the two sizes of specially designed ashtrays, the smaller's rim should be made to slope inwards; that the larger should be made smaller.

On 3 November *Oriana* was christened and launched by H.R.H. Princess Alexandra. No fitting-out of the public rooms had as yet been started on board.

November 1959–November 1960

The artists were now being commissioned for their murals and reliefs, and for the perspectives required for the various publicity brochures that would be required to accompany *Oriana*'s maiden voyage. Completion date for the ship was 21 September 1960 and inevitably the pressures began to build up. Critical points become apparent and the first notes of alarm may be detected. The architect's drawings have not been received by the contractors for two veranda bars. In April, as the names of the public rooms are finally fixed ('Men's Hairdressing' for tourist, but 'Barber' for first class), overtime is mentioned for the first time. There is a plumbers' strike; the fixed seating in one of the restaurants is held up for three weeks while the flooring is completed; please can outstanding information from the architects be had in sketch form if not in fully detailed drawings?; there is a big hold-up in the alloy doors and those not received may have to be replaced with timber; shortages of labour are making themselves felt. The completion date is moved from 21 September to 15 October. The cinema ceiling height is lowered to clear girders and consequently affects the projector beam; 60 alloy doors are still outstanding; testing of the air-conditioning units and important electrical equipment is made impossible by delays in the turbo-generator trials. H.M. the Queen is to inspect the ship on 21 October. The owners decide to hold the vessel at Barrow until 3 November, to ease the strain on the pursers in preparing the ship for sailing. 62,500 pieces of cutlery and 24,000 glasses are delivered. The design of the lectern and hymn board is approved. 1,436 electrical fittings are still outstanding. The laundry machines need lowering. Shall 50 per cent of the furniture be delivered to Falmouth? The planting contractor gives personal instruction to ship's personnel on the care and treatment of plants.

On 5 November Ship No. 1061, *Oriana,* finally left Barrow, having been held up for two days by bad weather. She sailed down to Falmouth for dry-docking for a week, before going to the Clyde for acceptance trials and then coming south once more. On 3 December, the embodiment of 6,500 working drawings by naval architects, engineers and interior designers, she sailed on her maiden voyage, outward bound from Southampton for Gibraltar, Naples, Suez and points east to Sydney, Auckland, Honolulu, San Francisco and Los Angeles. By the time she docked again at Southampton, fifteen weeks later, she had carried 10,648 passengers.

Final touches were being carried out in the public rooms until she sailed and a considerable list of minor items needed further attention at the end of that voyage and several subsequent voyages. These items were of no real significance, however. The owners expressed themselves well satisfied. It had been their intention, in setting up the design machinery they did four-and-a-half years earlier, to achieve variety of expression and adherence to a given completion date. 'This', they state, 'was to a large extent achieved and relieved the owners of much anxious and time-consuming work that they would not have been equipped to undertake.'

10. Entertainment

Theatre, film and television combine to an extraordinary degree the simultaneity of sport and music with the forward planning and problem-solving of the designer. In them, many creative talents are brought to bear upon the form of the production under a group leader: the producer or director. They are faced by innumerable physical and other limitations which must all the time be taken into account: floor space, sight lines, the nature of the particular equipment employed, the weather, the capabilities of particular performers, the budget, the need for split-second timing and a score of similar considerations.

After all the preplanning that such things demand, the challenge of the performance itself remains to be met. However long the run of the play, its performance can never become completely routine but remains subject to a multitude of small adjustments arising from the subtly changing relationships, deliberate and accidental, of the players, and indeed the audience.

Television provides the most striking example of this kind of teamwork. Here the achievement of a particular effect, a moment of exact significance upon the screen, is not only the summation of pre-planning by many skills but of acute sensitivity and immediate response by twenty, thirty, fifty or more different people simultaneously.

To analyse in any depth the creative processes underlying so intense a form of collaboration would take a book in itself. What follows is a brief examination of the role in that collaboration of one member of the team, the television designer. His ancestors—the Bibienas, the Berninis, the Inigo Joneses—could call upon the resources of court and nobility for the creation of their masques and spectacles. He may be called upon to design as many productions in a year as they designed in a lifetime, and to back him must have resources no less considerable.

The largest television design department in the world is to be found within the British Broadcasting Corporation. Size as such holds no particular virtue. It may, however, be of interest to see how creative energy is harnessed and co-ordinated within so large an organization. In the terminology of our theme, the BBC's design and allied service departments may be regarded as a form of integrated practice, operating through largely autonomous groups assembled for each production. Although the resulting programme is a one-off product, it is, so to speak, designed for quantity reproduction and the size of the television audience raises all the aesthetic and moral considerations that are inevitably bound up with mass communication and entertainment.

Accidents of history and English empiricism have combined to give the British Broadcasting Corporation its form. Like, say, the Tate Gallery, it embraces functions which more rational countries have seen no reason to combine under the same direction. The BBC is responsible for three distinct services, each one of which constitutes a sizeable concern. Within Britain, it originates and transmits around 380 hours of programmes weekly for three domestic radio channels and a network of regions. (It remains to be seen how the new purely local radio stations fare.) It is similarly responsible for about 100 hours weekly on two television channels, likewise networked through regional services with their own variations.* It is further responsible for over 660 hours of external broadcasting beamed, in over 40 languages, to all parts of the world. Though now about sixth among the nations in terms of output, continually threatened with further cuts and weakened by low powered transmitters, the Corporation's external services, it may be noted, are re-broadcast by more stations than are those of any other country.

The Corporation is also very English in its quasi-official character. In company with such bodies as the British Council and the Council of Industrial Design, the nationalized industries and a host of Royal Commissions and Councils, it occupies that territory, so mystifying to foreigners, which looks as though it should be subject to Government control, or at least intervention, but to all intents and purposes is not. The BBC is a non-profit-making body corporate, set up in 1927 by Royal Charter (before which, for five years, it existed as the British Broadcasting Company). It derives its funds, now in the neighbourhood of £65 million a year, in the main from the mandatory licence fees of 16 million viewers and listeners, which are collected and transmitted to the Corporation by the Postmaster-General. In respect of its external services, the Corporation receives a grant-in-aid of some £10½ million a year from the Treasury and the extent of these services is partially shaped to meet the wishes of the Foreign and Commonwealth Offices. Programme content, however, remains the responsibility of the BBC (as was evidenced, for example, during the Suez crisis); here, as in its home services, the Corporation guards jealously its autonomy of action, freedom of speech and the independence of its programmes from political, commercial or other pressures.

From its first Director-General the Corporation inherited a powerful streak of nonconformist puritanism: always a well-nigh ineradicable trait in any family. 'Nation shall speak peace unto nation' was the Corporation's motto until 1939 killed it. To 'entertain, inform and *instruct*' (my italics)

* By contrast, no fewer than fifteen companies network their programmes to produce the United Kingdom's single commercial television channel.

summarizes the terms of reference enshrined in the BBC's Charter. The Corporation's Olympian calm received a rude jolt, however, with the coming, in 1954, of a second TV channel operated through a network of commercial companies under the general aegis of the Independent Television Authority. Under the unwonted spur of competition, programmes became brisker, harsher and more irreverent. A generation schooled in the Reithian concept of public service was pushed aside by young men who came to television as an accepted means of communication, rather than a marvel. On the screen things were said and done which, unthinkable even a decade before, stung organized protest groups into being. Nonetheless, notwithstanding the pressure to hit and maintain high audience figures, the BBC continues in general to exhibit qualities of authority and responsibility which are the envy of most other countries.

The Corporation is controlled by a Board of part-time Governors (currently nine) normally appointed for a five-year term. Three of these specifically represent Scotland, Wales and Northern Ireland. Around the Board is grouped a multitude of committees or councils, the broad function of which is to ensure the representation of balanced political, religious, local and specialized views. Scottish and Welsh Broadcasting Councils assume responsibility for the content of programmes in those areas; their members are nominated by the, in 1965, 63-strong General Advisory Council. There are, in addition, Advisory Councils for the regions, for schools broadcasting, for religion, science, agriculture, appeals and so. Armed thus with the views of the nation the executive head of the Corporation, the Director-General, reports to his Board of Governors. He is accorded, it may be assumed, the freedom and autonomy normal to the chief officer of an organization employing over 21,000 full-time staff, with an annual budget considerably greater than that of Luxemburg.

From the Director-General the chain of control fans out through a Board of Management, composed of the Director of Sound Broadcasting; the the Director of Television Broadcasting, the Director of External Broadcasting; the Director of Engineering, the Director of Administration (responsible for finance; staff appointments, training and welfare; accommodation and central services; all legal matters, including programme contracts and copyright) and the Chief Assistant to the Director-General (responsible for all those items which do not fall neatly into other spheres of action, such as audience research, publications and overseas relations).

Each member of the Board of Management heads a considerable organization. As the balance between sound radio and television has shifted, however, the latter has come to dominate the established pattern of Corporation affairs. Of the 21,500 employed by the BBC in 1965, over 13,000 worked in television. Of the £40 million a year operating expenditure in the mid-sixties, £27½ million was swallowed by television. Expressed in direct relation to output, the cost of television compares even more dramatically with that of the older services of the Corporation. On average, an hour of external broadcasting costs £217; an hour of home broadcasting costs £704; an hour of television broadcasting costs £5,465.

It was the BBC which, in November 1936, from Alexandra Palace in north London, inaugurated the first public television service in the world. Today Alexandra Palace is still in use but primarily by the news services, which operate independently of the rest of the organization and have a direct link to the Director-General. In all other respects BBC television has gravitated to west London. A scatter of studios stretches from a theatre on Shepherd's Bush Green, through Lime Grove—for long the main centre—to the Riverside Studios and the old Ealing Film Studios. All now revolves, however, around the Corporation's circular Television Centre, purpose-built in an unlovely area between Shepherd's Bush and the White City Stadium. Here is the core of the empire over which the Director of Television Broadcasting presides. Here is to be found the Television Design Department.

Those directly concerned with the creation of programmes fall into two broad categories. There are firstly those who originate and produce the programmes: in the designer's sense, these are the 'client' departments. Secondly there are those who provide these departments with a variety of services, including a comprehensive design service. The former are currently grouped, under a Controller responsible to the Director of Television Broadcasting, in thirteen Programme Departments: Planning, Presentation, Drama, Light Entertainment, Outside Broadcasts, Current Affairs, Music and Arts, Documentary, Travel and Feature, Further Education, Family Programmes, School Broadcasting and Religious Broadcasting. The latter, under a second Controller, are currently grouped in eight Programme Servicing Departments: Films, Scenic Servicing, Design, Studio Management, Costume, Make-up, Artists' Bookings and a Script Unit.

As one approaches closer to the moment of truth, represented by performance and transmission, the deployment of an ever-growing range of specialized skills becomes increasingly apparent. Television is a team activity of great complexity. It calls for the most sensitive collaboration between scriptwriters, actors and other performers, composers and executive musicians, sound engineers, lighting engineers, camera crews, scene painters, costume designers, make-up artists and a multitude of technicians and production assistants; quite apart from the logistic support of administrators, accountants, drivers, porters, maintenance and supply men. Design for television is perhaps the most intensive form of multi-discipline group practice that exists.

The Design Department works upon 75 to 100 shows a week. These are of three broad types: Talks (news, discussion, documentaries and panel game programmes); Drama; Variety and Light Entertainment. The type of

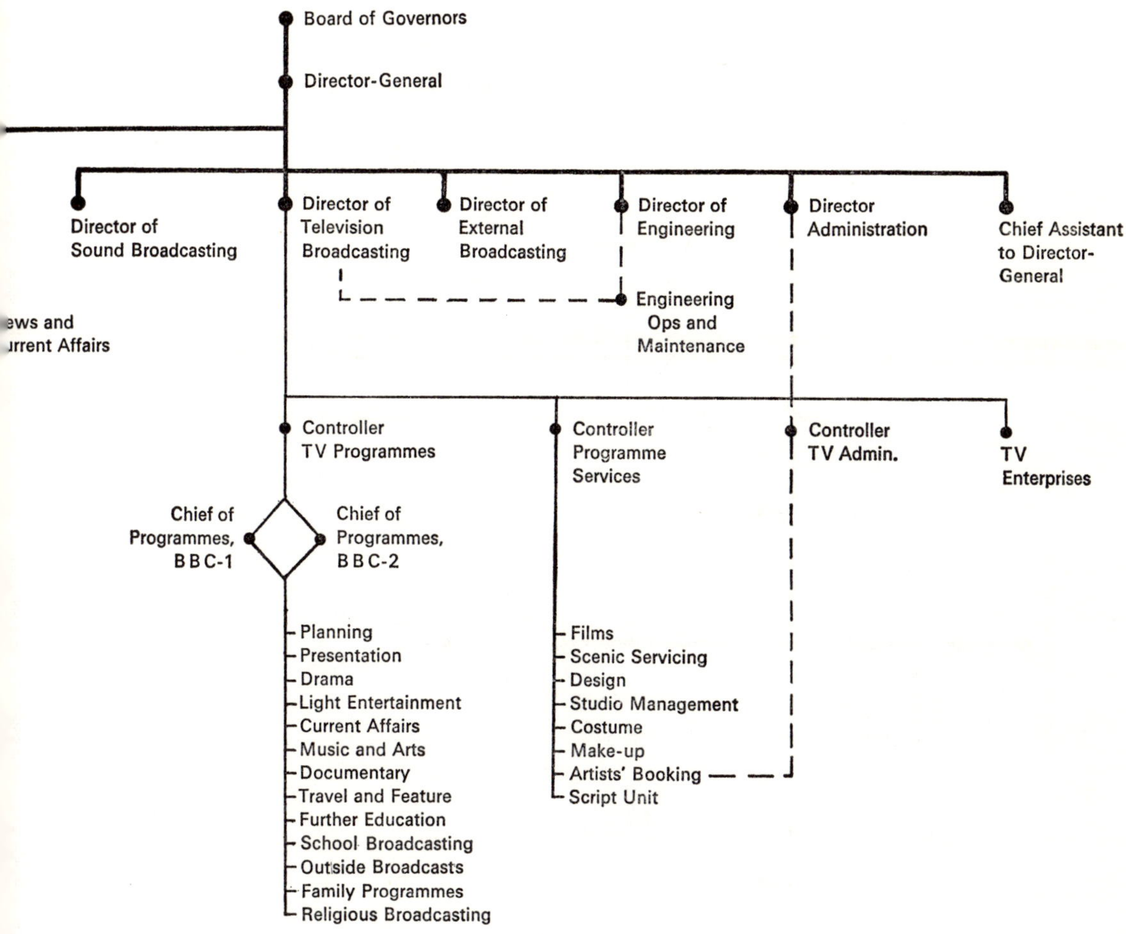

British Broadcasting Corporation: general organization.

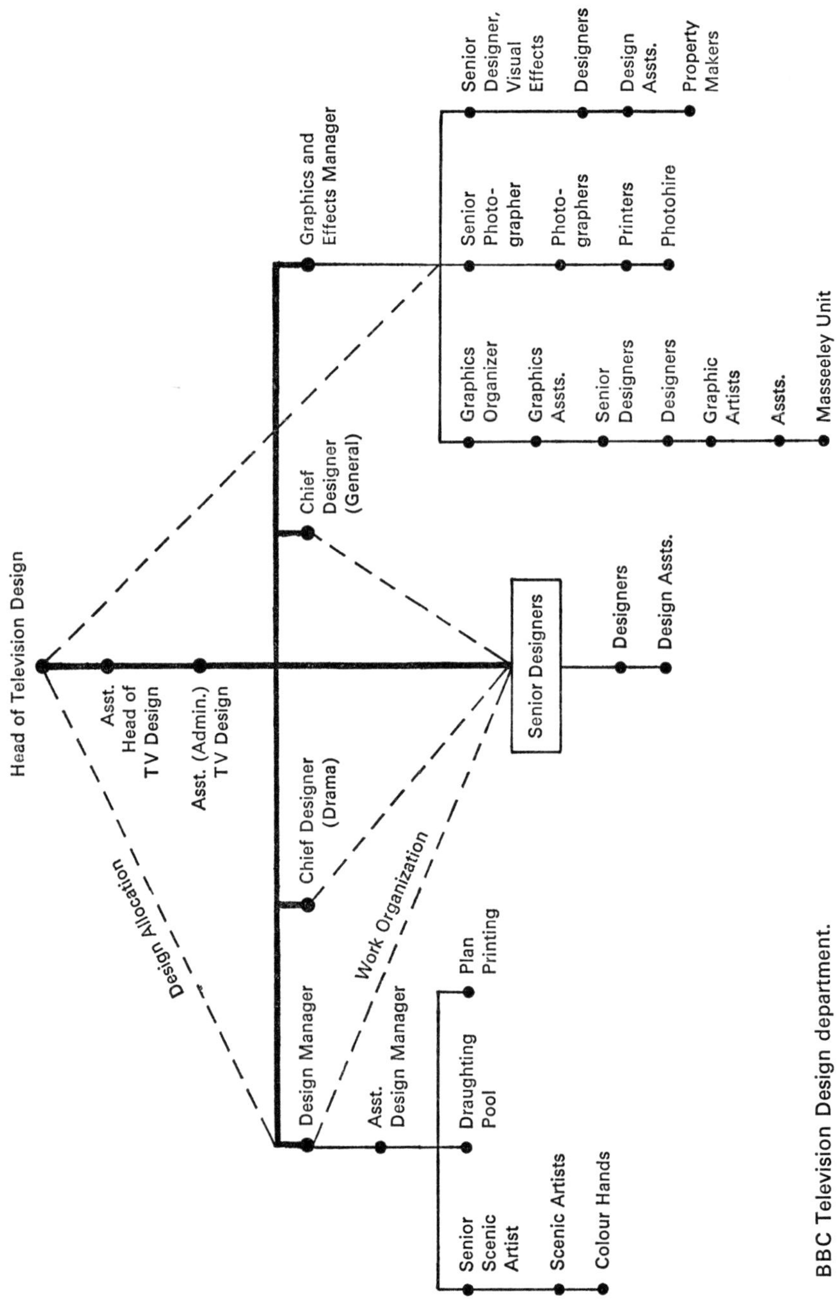

BBC Television Design department.

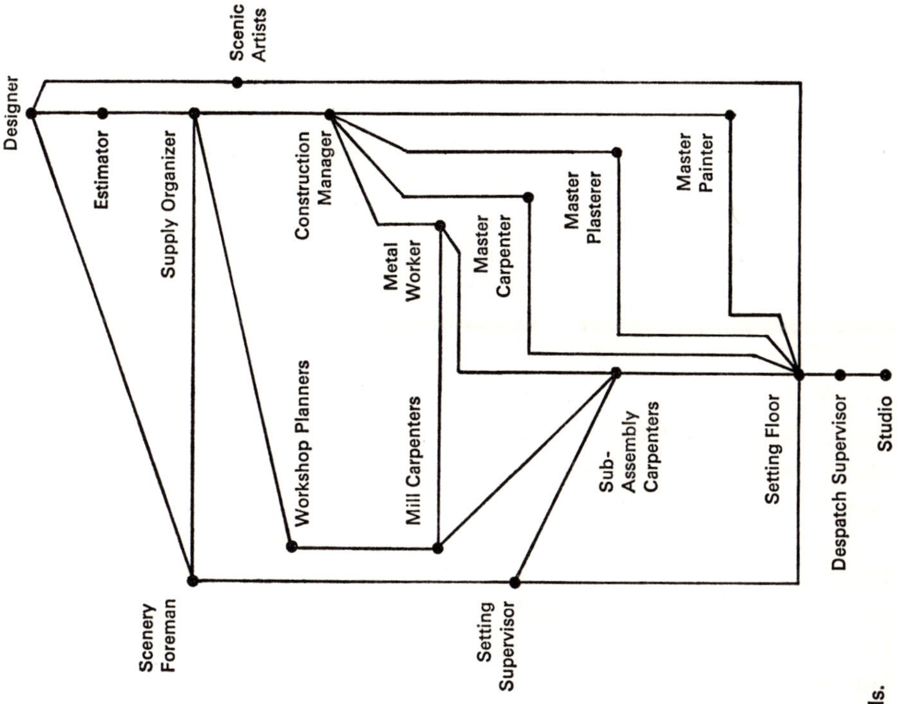

Realizing the design of a television set: a network of skills.

visual framework demanded by these programmes varies greatly, from the wholly realistic, for many kinds of drama, to the formalized, the wholly abstract, and what may be termed simply the 'studio interior' required for, say, a political discussion. Special effects of every kind are regularly called for: rain, fog, fire and snow; photographic montages and back-projection; things from outer space, ships at sea or models of the molecular structure of ribonucleic acid. No less than 10 per cent of all transmission time is taken up by graphics: titles, maps and diagrams, still and animated. All these things have to be designed. There is no such thing as a neutral image on the screen, save by design, and it is the designer's task to make the visual framework of each changing image as meaningful as possible.

The BBC's Television Design Department dates from 1953. In that year Richard Levin, who for twenty years previously had had contact with the Corporation as an exhibition designer in private practice, joined the Corporation to head a Department of Design and Supply. Facilities were minimal. A dozen years later, with Levin still at its head, the Department consisted of 268. Some 65 of these were designers of varying seniority, aided by 48 Design Assistants. To the other creative workers—photographers, model-makers, scenic artists—was added the essential support force of administrators, cost accountants, estimators, secretaries and the rest. A small number of designers were attached to the regional centres. In addition to its whole-time staff, the Department employs up to 10 freelance designers, primarily to meet unavoidable peaks in the work load but also to bring particular talents to bear upon appropriate programmes. A certain number of hours, at a standard hourly rate, is guaranteed to such designers during their first year in private practice.

With its great increase in size, the Department has inevitably grown more impersonal, its procedures governed by more remote forms of control. No longer is it possible for a designer to roam through props with a shopping basket. He will never find himself mucking in, like his theatrical counterpart, mixing paint for a backdrop. Although Levin had declined to take on 'supply' at the very outset, it was only in 1963, with the expansion occasioned by the coming of the second BBC channel, that the workshop side of the operation was hived off to form the Scenic Servicing Department—contractors, as it were, to the designers.

The Department now channels its skills into three main streams. Under two Chief Designers, one for drama and one for general programmes, come thirteen senior designers, each of whom heads a small group of about four designers and four design assistants. A second stream, under a Design Manager, includes scenic artists and colour hands, a draughting pool and a plan printing unit. The third stream, under a Graphics and Effects Manager, embraces all graphic designers, photographers, darkroom assistants, special effect designers and property makers. The effort represented by all these

ges on the screen: fragments only from Episode 3 ostoievsky's *The Idiot*, transmitted on BBC-2, but aps in them may be glimpsed something of the qualities which distinguished this production—a nostalgic bloom, an integrity of atmosphere and setting, an exceptional richness of lighting.

—G.P.D.

nages in the studio: space, night, mystery . . . but the tudio walls, above, are only inches away and the tors, below, have only a few square feet in which to play some of their scenes. This is the way things looked during and between camera rehearsals, before the play was recorded.

sets all pivot about this great tree and lead
rceptibly one into another, so that each provides
ı and authenticity to the others. There is half a day
› yet, before recording: the fallen leaves of autumn
—three sackfuls—have yet to be scattered about the ground, mikes have yet to be moved, sunlight and moonlight have yet to dapple the grass. Yet the validity of this extraordinary 'exterior' can already be felt.

With a bandstand the set forms a park, above; beyond the verandah of Lebediev's villa, it becomes a garden, below. The same verandah, seen from another angle, top right, becomes a wing of Epanchin villa; from inside, bottom right, the sense of depth and space is vast. In reality, it is shorter than a cricket pitch

Here, on the gridded floor plan of Riverside 1, can be seen the tight-knit ingenuity with which the designer, Tony Abbott, used the limited space at his disposal. Dimensions are 68 ft. by 64 ft.; printed round the sides are indications of available services; pencilled in, as circles over the setting lines of the actual scenery, are the director's chosen camera positions—evolved notionally, in consultation with the designer, weeks and months before the sets come into existence. They will be revised and adjusted on the floor of the studio, to take advantage of freshly realized possibilities, but this is the document about which all preparations for the production turn—actors' moves in distant rehearsal rooms, scenery construction in workshops, the hiring and purchase of props, provisional lighting plans and a multiplicity of detailed decisions. Once finalized, the designer's floor plan embodies the future success or failure of the production; if he has failed to visualize the results correctly, nothing afterwards can change the fact.

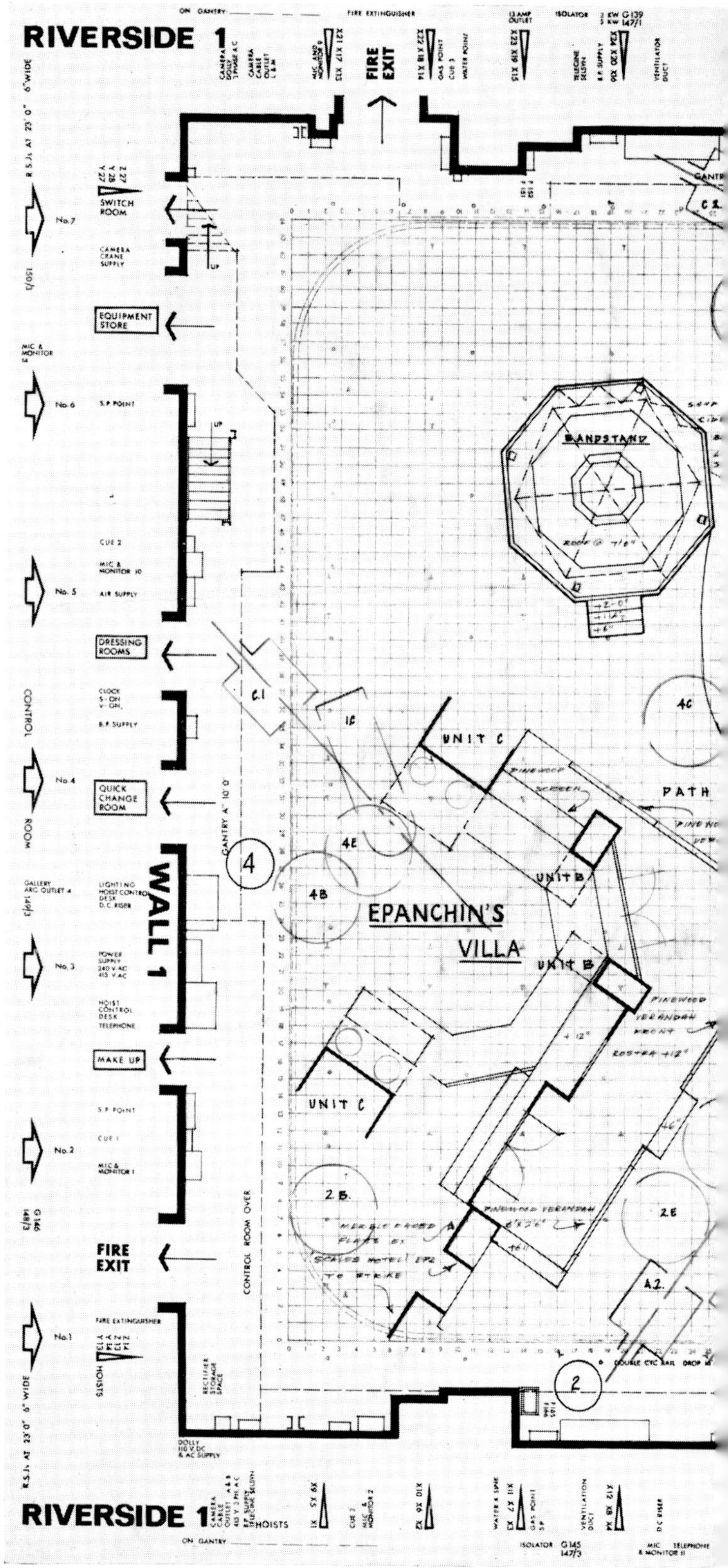

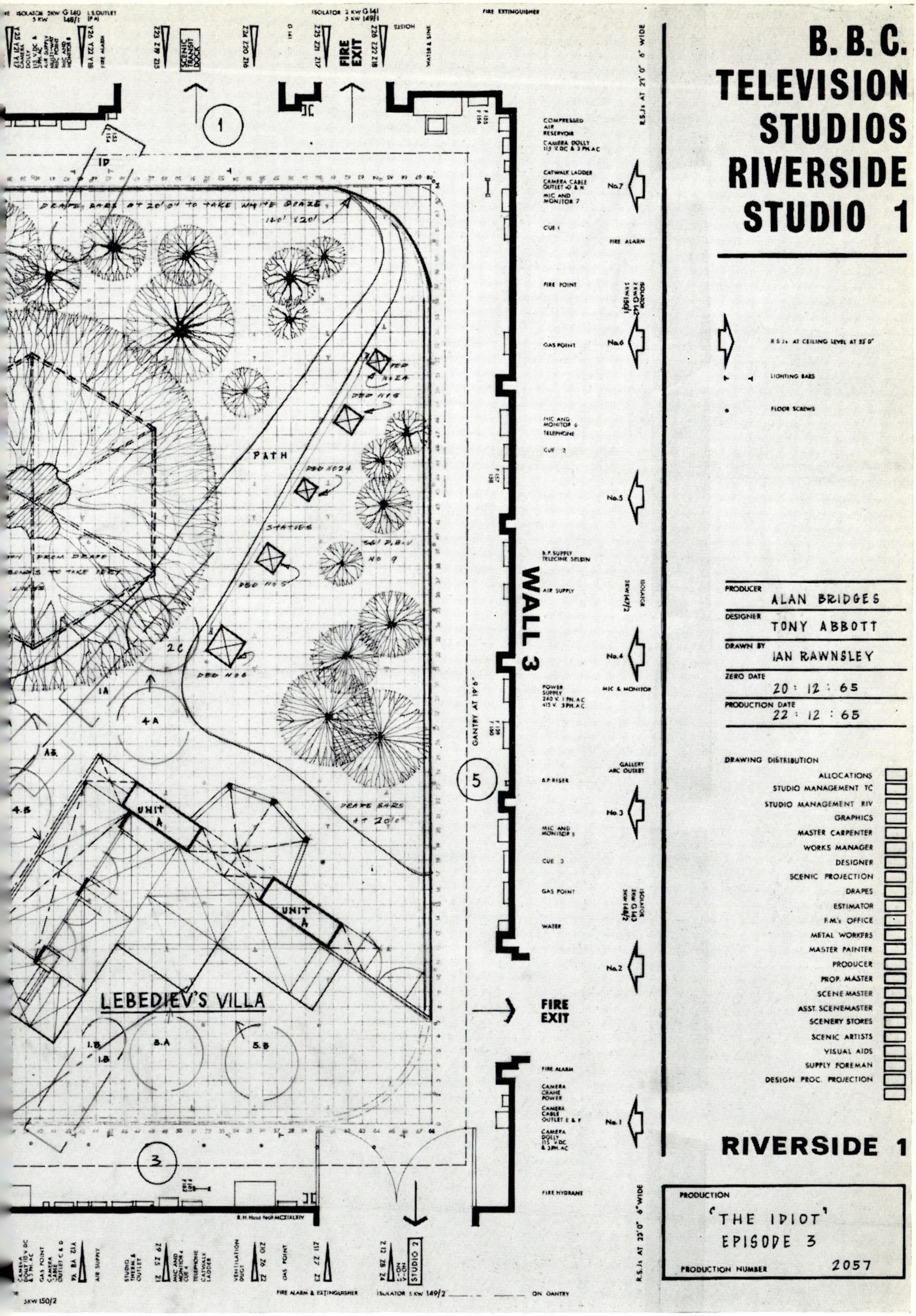
B. B. C.
TELEVISION
STUDIOS
RIVERSIDE
STUDIO 1
R.S.J.s AT CEILING LEVEL AT 33'0"
LIGHTING BARS
FLOOR SCREWS
PRODUCER
ALAN BRIDGES
DESIGNER
TONY ABBOTT
DRAWN BY
IAN RAWNSLEY
ZERO DATE
20 : 12 : 65
PRODUCTION DATE
22 : 12 : 65
DRAWING DISTRIBUTION
ALLOCATIONS
STUDIO MANAGEMENT TC
STUDIO MANAGEMENT RIV
GRAPHICS
MASTER CARPENTER
WORKS MANAGER
DESIGNER
SCENIC PROJECTION
DRAPES
ESTIMATOR
F.M.'s OFFICE
METAL WORKERS
MASTER PAINTER
PRODUCER
PROP. MASTER
SCENE MASTER
ASST. SCENEMASTER
SCENERY STORES
SCENIC ARTISTS
VISUAL AIDS
SUPPLY FOREMAN
DESIGN PROC. PROJECTION
RIVERSIDE 1
PRODUCTION
'THE IDIOT'
EPISODE 3
PRODUCTION NUMBER
2057
WALL 3
FIRE EXIT
FIRE EXIT
PATH
STATUES
UNIT A
UNIT A
LEBEDIEV'S VILLA
GANTRY AT 19'6"
DRAPE BARS AT 20'0"
R.S.J.s AT 23'0" 6" WIDE
R.S.J.s AT 23'0" 6" WIDE
No.7
No.6
No.5
No.4
No.3
No.2
No.1
STUDIO 2

LEE SPILLER
12'0"
4'0 APPROX

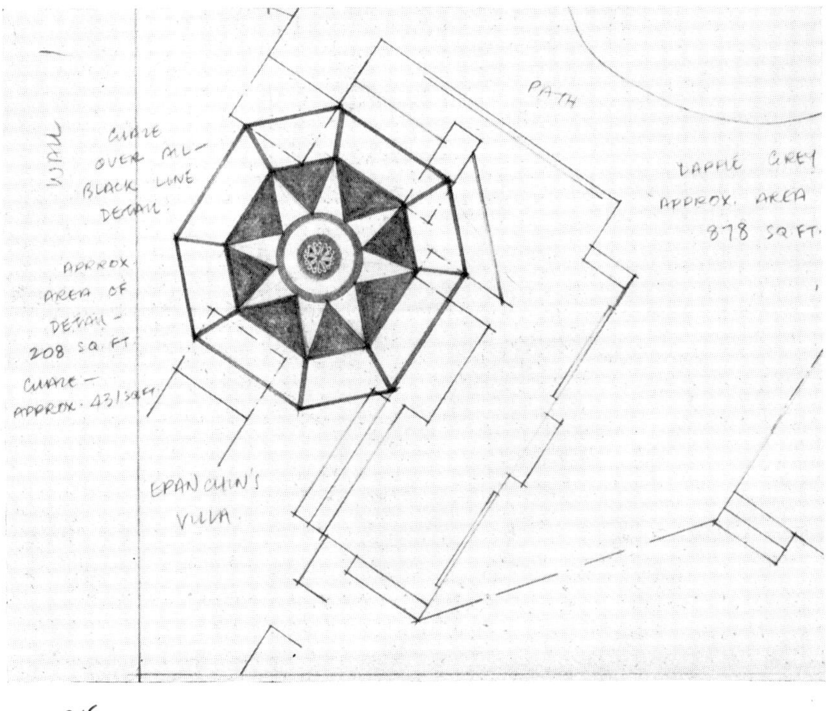

WALL
PATH
APPROX AREA OF DETAIL - 208 SQ. FT.
APPROX. AREA 878 SQ. FT.
EPANCHIN'S VILLA

r the five episodes of *The Idiot*, Tony Abbott pre-red forty constructional drawings, plus other detail awings and plans. Since he was to hire a good deal well-detailed film scenery, these drawings for the ost part were very simple and related to stock units nich could be used throughout the production. Less ual was the elevation of the proposed tree trunk, left. pical, on the other hand, was the detail for floor inting, below left, in Epanchin's villa (it can be seen the foreground of the completed set at the foot of ge 243). Two of the uses to which Abbott put his ecially constructed elements can be seen in studio ots, right and below, from other episodes.

More studio shots of scenes from other episodes of *The Idiot*. From simple elements like those on the previous page, from the rich detailing of floors and furniture and props, from lighting that is always crystal clear about what it wants to reveal and what it can afford to lose, can be compounded that fragile some thing which causes the imagination to take wing.

specialists is allocated and focused through the Design Manager, in consultation with the Head of Design. He it is who sits in with the programme planners, reads advance scripts and evaluates, in relation to the capacity of the Department and the scenery workshops, the resources to be made available to any given production.

The budget is clearly a major factor in shaping any programme. Until fairly recently the Departments concerned made their own estimates of the resources they could allocate to a particular production, while keeping within their own departmental budgets. Gentlemanly disputes and blackmail were the inevitable result. The dialogue went something like this. Director to Designer: 'Can't you manage something a bit more ambitious?' Designer: 'But I've had to cut it to the bone as it is.' Director to Planner, or even Controller: 'I've spent £*x* thousand already on a couple of stars—we can't spoil the ship for a ha'porth of tar.' Planner, wearily: 'I'll authorize another 200 man-hours.' Collapse of Design's departmental budget. As a result of such confusions, Producers are now required to make their own PBE, or programme budget estimate, at the outset, in consultation with the Service Departments concerned. Since they themselves will be held responsible, should the figure later seem oppressively inadequate, they are unlikely grossly to underestimate in the first place; on the other hand the normal play of supply and demand operates at the planning level since the planners have to work to *their* own budget. The average cost of programmes is well known and inflated budgetary demands by a director will only price him out of the market.

When the programme commitment has been agreed in a form acceptable to the Design Manager, he submits his estimates and suggestions to the Head of Design; together they allocate specific designers to the programme in question. When a week's work has been agreed in this way, the information is transferred to the Design Manager's wall chart, which shows him the commitments of all designers over an eight-week period, and is circulated throughout the Departments concerned by means of a stencilled bulletin.

The design groups thus formed normally consist of four people: the designer, an assistant, a draughtsman and a graphics man. The time and resources available for the job are clearly defined in advance in terms of man-hours.* The allocation naturally varies from production to production —a variety designer can turn out maybe ten programmes to one from a drama designer—but on average each set represents 300 man-hours of construction, 100 man-hours of organization and supply of properties, a day and a half in the studio and ten days' work by the designer and his assistant. Since the whole process of bringing a production to the screen lasts about four

* This system sometimes militates against flexibility. A designer once tried to 'sell' 300 man-hours for cash with which to construct or purchase props. The administrators would only give him half the 'face value' of his man-hours.

weeks from the designer's first discussions with his producer, or director in the case of drama, it will be seen that the television designer usually has several programmes in the pipeline at any given moment.

Notwithstanding the powerful machinery that has been evolved to aid him, the pressures under which he works are often appalling. Design for television partakes to a considerable degree both of the traditional processes of theatre design and of the newer traditions of the cinema. In these, however, the designer is seldom concerned with more than one production at a time: a more leisurely overall tempo and fewer technical considerations in the one, an absence of absolute deadlines and the chance to reshoot repeatedly in the other, make it possible to qualify or amend first thoughts. Not so with television. Such is the flow of programme material required by this insatiable medium that time permits the designer no trial-and-error approach. Not only must every preplanned detail fall perfectly into place during the brief 24 to 48 hours that the production is in the studio; within another round of the clock that set will have been broken up and he will be immersed in the initial problems of another production; probably with a third, on which he has been working, coming up for transmission within the fortnight.

It is said to take about two years' apprenticeship before a designer fully grasps the techniques of television. Small wonder that by then some are already exhausted creatively. The very size of the television audience prohibits the re-use of ideas. A single performance of *Salome* on BBC Television reached a larger audience in one night than all previous performances totalled over the half-century of the opera's life. Just as the music-hall comedian can no longer exist for years on a single set of jokes, so the designer has to produce 'a flow of new ideas for a situation which is basically always the same'.* The maintenance of standards becomes a problem in its own right.

The BBC has a 'quality committee' which samples programmes at random and reports on the technical proficiency of the sound, the lighting, the camera work and so on.

> In a major television service, [Richard Levin has written, however], it is impossible for one man . . . to attempt to keep personal control over the whole visual output, if only by reason of the mere quantity of it: effective delegation of responsibility for maintaining an aesthetic standard is therefore essential . . . a formula must be found for dividing responsibility . . . between groups of specialists concerned with particular categories of design . . .
>
> Design groups should be kept small enough for a senior designer to control and administer without difficulty . . . Group working of this sort promotes discussion and a lively interest in the work of all its members. It also encourages an intensive and valuable self-critical faculty which has little time to operate during the ordinary course of working—unless there

* *Television by Design*. Richard Levin. Bodley Head, 1961.

is deliberate encouragement for it in this way. Having established a design group structure under the control of the senior designers, and having promoted active discussion of design practice at this level, discussion and consultation between the senior designers in charge of groups will provide a valuable cross-reference between all the interests of a design department.

In practice, such design policy meetings of Senior Designers are held monthly under Levin's chairmanship.

This emphasis on group working is in line with current BBC policy generally. The simple authoritarianism of the Corporation's pioneer days belonged to another age; it seems light-years away from the activities of the bright young men who emerged during the early nineteen-sixties from the tutelage of the indomitable Grace Wyndham Goldie of Television Talks. Decisions, the Director of Administration stated recently, should be taken as near to the actual operations as is consistent with the overall aims of broadcasting. The best level for every decision, he added, was usually a good deal lower than was generally thought.

On today's scale of operations, broadcasting generally, and television pre-eminently, demand such interlocked preplanning by such a multitude of skills that it would be difficult to run a service on any other basis. A misjudgement by any one of the team during preplanning may easily start a chain reaction of adjustments that builds up during performance to the point of complete disruption. A yard difference in the placing of a table may block the track of a camera dolly; the re-alignment of the camera perhaps brings into view a quick-change booth off stage, which, moved to a new position, adds several seconds to an actor's re-entry, geared to which is a pre-recorded music track. The opportunities for upsetting the whole precarious balancing act are so manifold that disaster would be endemic, did not the entire production team identify itself personally with the success of the programme.

The producer (or director in the case of drama) is the creative pivot around which the whole operation revolves. It is he who must not only set the tenor and character of the programme but ensure that all the specialist skills around him are fully in the picture and in accord, at least from a technical point of view, with the way it is developing. Director and designer discuss the programme at the outset: this is indeed the first creative act in the process of making a programme, once the artists have been booked, the programme scheduled and a studio allocated. Conferences and discussions thereafter are well-nigh continuous. Most of these are *ad hoc* but at an early stage a main planning meeting takes place, attended by sound and lighting engineers, production and costume designers, and everyone else primarily involved, specifically to co-ordinate all groups and departments.

Such meetings are informal, with a good deal of give and take. A sound engineer expresses doubts about the producer's intention to give a singer a hand mike, not on grounds of technical difficulty but because the artist is an opera singer and it seems out of character. The designer points out that if a studio door were taken off its hinges, the orchestra could be accommodated differently and such-and-such a camera angle would thereby be made possible. And so on. It is all very friendly but, for the most part, design is seen by all but the designers themselves as functional rather than visual. Almost without exception, producers and directors are ex-actors; it may be said without disrespect that few actors are visually orientated. Organizationally this attitude is reflected in the looseness of relations between the production designer, the graphics designer and the costume and make-up artists. Graphics, for the most part excellent, are rarely integrated fully into the fabric of the programme as a whole; make-up is usually impeccable but it becomes clear as soon as a production moves into untraditional fields, such as science fiction, that costumes are conceived in perfunctory relationship to the settings. This is a position which must change with the coming of colour.

Functions, on the other hand, are clearly defined and friction on this score is reduced to a minimum. Where disagreements cannot be resolved—disputes on facilities, on the short time allowed by a previous production for setting the studio, the designer's man-hour allocation, artistic disputes with the producer—these may have to be passed upwards, to the Head of the Programme Group and the Head of Design or other Service Group, even the Controller, for decision. In practice, such situations tend rather to produce grumbles *after* the show has gone out, for the simple fact is that time does not permit the luxury of formal disagreement before that.

It will be seen then that, in television, the design process is conditioned by a myriad of factors outside the control of the designer. Pure television design as an autonomous concept, in the sense of the theatre designer's total statement, does not exist. Rather is it an additive process, concerned with the provision of fragmentary images on the viewer's screen, the cumulative effect of which will set undertones and overtones of meaning reverberating in the viewer's mind.

Just how the designer contributes to those images may be judged by studying a particular production in greater detail.

On five Tuesday evenings, between 11 January 1966 and 8 February 1966, a serial dramatization of Dostoievsky's *The Idiot* was broadcast by the BBC on its second television channel, BBC-2. The production, notable for its integrity and for its sense of place and period, was something of a *tour de force* visually. It received a higher than average 'reaction index' from the BBC's Audience Research department. Of it the *Guardian* said: 'If all the good work of the BBC in drama could be totted up, for sheer quality its sustained production of serials from classic novels would have to come near the top.' *The Times* was using the production as a yardstick several months later. For his work in 1966, the designer, Tony Abbott, received the Guild of Television Producers and Directors Award, 'Designer of the Year'.

Altogether over sixty people, excluding the cast and much support labour, were directly concerned with getting the show on the air. Long before most of them began to contribute their special skills to its realization, however—indeed long before most of them had any idea even that the title was scheduled—essential decisions were being taken which were forcing the production into a certain form and structure. The story began, to all intents and purposes, in the second quarter of 1965 when Douglas Allen, a classical drama producer, 'made an offer' of *The Idiot* to the programme planners of BBC-2 for production in the last quarter of the year. For programme purposes BBC Television divides the year into quarters, a separate planning unit being responsible (in consultation with the Head of Programmes for the channel concerned) for the co-ordination, balance and budgeting of all programmes within that period (which tends in consequence to be known by the name of the planner concerned, as So-and-so's Quarter). The Channel Head will have laid down the broad policy by which he wishes programmes to be shaped, bearing in mind the need for an internal balance of interest within the channel as a whole and to provide an appropriate alternative, in general character and in detailed timing, to other channels. Within this general policy, the Output Departments formulate their own proposals for offer to the planners at least six months before the quarter in which transmission would take place. Series and serials form the structural framework of programme planning. One such continuing series on BBC-2 was concerned with classical drama and thus it was that Allen, in the light of preceding titles and the various possibilities that suggested themselves, made his 'offer' of *The Idiot*. An earlier production of *The Brothers Karamazov* had proved highly successful;* it was felt that *The Idiot* was a natural successor.

By mid-June the idea had been accepted by the planner; by mid-July it had been scheduled for transmission in January; by August its recording dates had been fixed. During the second half of June, Allen turned his thought to script and production team. His first task was to commission an adaptation of Dostoevsky's text. For this he approached Leo Lehman, an experienced television scriptwriter. For director his mind turned naturally to Alan Bridges—in the words of the *Sunday Times* 'a veteran master of the genre'—who had already shown, in *Karamazov*, his ability to identify himself with the spirit of Dostoievsky. Bridges accepted the offer with pleasure and turned to the more immediate problems of a film he was making, *Murder by Accident*, and a play he was producing in the West End.

Alan Bridges, tall, fair, in his mid-forties, an ex-actor turned director who continues to work in the theatre and in films, had been in television for five

* It reached, said the *Guardian*, 'a new level of insight into character and sympathetic atmosphere'.

years. He could nonetheless be described as a second-generation director: one of those with a markedly more flexible approach to production than their predecessors of the first. In television's early days, rudimentary facilities had a powerful effect on style. Outside rehearsal rooms were primitive; studio space was limited; cameras were few; recording facilities and linking film sequences (with all the possibilities they open up of striking or redressing a set, of changes of costume or make-up by the cast) were almost non-existent. Disaster, which can never be far out of sight in television, loomed that much closer. The result was that production had to be pre-planned in every detail, to a point of iron rigidity, if the show was to be got on the air at all. It is still standard practice for directors, about two-thirds of the way through rehearsals, to work out their camera line-ups; abstract them into shot lists for each camera; and adhere to them through thick and thin in the studio. Among lighting men, camera crews and actors the attitude persists that positions thus notionally established during rehearsals must be retained in the studio and that to change them at this later stage betokens something near professional incompetence.

Increasingly, however, the tenets of the rule-book are adjusted by the style and intelligence of the individual director. Bridges and his like approach the moment of truth, the moment of transmission, with a more open mind. Once in the studio, with the realities of set and props and lighting before them, they are prepared to amend and rethink, to refine and refocus, to improvise and improve, right up to the moment the second hand reaches zero. Lighting is adjusted, or totally changed, to sharpen the particular emphasis desired. Cameras are repositioned so as to catch, by *exact* framing, not merely the story-line content of a shot but all its emotive undertones and overtones as well. Totally new shots are introduced into the script ten minutes before transmission; if as a result there is no backing to that pair of double doors, for Heaven's sake use the wall of the studio itself. Production on these lines inevitably makes greater demands upon valuable studio time and strains the organizational machine accordingly; crisis and panic are not infrequent. On the other hand the effect upon cast and production team is wholly stimulating. Nothing is routine, everyone is keyed to concert pitch, and the image on the flattened tube is fresher, more exact, more compelling.

With Bridges now built into the production, creative responsibility moved to him. Allen as producer would retain administrative and budgetary control over the programmes but their actual form, vitality and integrity would result from whatever Bridges could extract from, or pump into, the multifarious colleagues he had now to assemble and weld into a team. The programmes were as yet but shadowy in his mind. Their reality resided in five duplicated scripts, of about fifty pages each, which were now to hand though not yet finalized; in five dates on a memorandum and in his diary; and in a PBE or programme budget estimate which, once accepted, would provide the financial limits beyond which he must not stray.

By August 1965 the total PBE had been agreed for the five episodes of *The Idiot* (now pinned down as Project Numbers 2324/3179-3183). It was perhaps a shade higher than for an equivalent production on BBC-1 but a good deal lower than a top-rating series. Alan Bridges's next concern was to choose his cast and get it under contract. He had by now re-read the original book by Dostoevsky and had formed a fairly clear mental picture of the characters as he would wish them played. Actors' diaries tend to read rather like railway timetables as their owners shuttle from theatre to Sunday poetry reading to film studio to personal appearance to regional broadcast. No producer is likely to get all the actors he wants. Bridges was fortunate, however; one first choice was in America but after much thumbing through of *Spotlight*, after consultation with Allen, after Bookings had applied themselves steadily to the actors' agents and in some cases the actors themselves, he had a capable cast under contract. They would meet for the first time early in November. David Buck would be playing the Christlike 'idiot'; Adrienne Corri would play Nastasia.

By September it became necessary to clinch the key members of his production team. When he had first taken on the production, his thoughts had turned, not unnaturally, to those who had worked with him on *Karamazov*. His designer then had been Tony Abbott, his lighting man James Maiden. Abbott, a senior designer, would be free in November and December and, although the allocation of designers is the responsibility of the Design Department, Bridges had early obtained Levin's agreement to make Abbott available. Maiden, based in Scotland, could not be used; however, another brilliant lighting artist, Robert Wright, who had been with the BBC since 1944 and with whom Abbott had worked before, could be brought in instead. This, too, Bridges contrived to clinch with Technical Operations, the department concerned. Thus was the basis for *The Idiot*'s superlative visual quality laid.

Tony Abbott had been looking forward to this job ever since he had first heard that it was on the stocks. He himself had trained as an architect, had been with the BBC for ten years; first in light entertainment and talks, latterly in drama, and had long admired Bridges's work greatly. Their first encounter, in fact, had been when he stopped Bridges in a corridor one day to tell him so, and to say how much he would welcome the opportunity to work with him if ever the chance arose. *Karamazov* had been mounted in Glasgow and, the production team having spent the best part of a month in the friendlier, more close-knit, atmosphere of one of the regions, the two men had come to appreciate each other's mode of working. Now, in early October 1965, both of them with other productions currently claiming the greater part of their minds—Bridges his film and Abbott the sets for Orwell's *1984*—they were meeting again to discuss *The Idiot*.

The initial meeting of director and designer is significant. The exchange of ideas which then takes place is really the first creative step in the long process of television production. Both will have studied the script. In their minds images, feelings, fragments of action, memories of the way sunlight fell once long ago, will have begun to form. These initial apprehensions are easily lost in the hurly-burly of detail that must follow. They frequently represent, however, the purest distillation of the spirit of the script; the degree to which the production finally succeeds more often than not reflects the extent to which they have been retained right through into transmission.

All that was so far firm was the pile of five scripts, each of which indicated in minimal terms the sets required—Train Compartment, Int. Lebediev's Hovel, Int. Rogojin's Corridor/Hall, Bandstand in the Park and so on, to a total of 29 in all. Known, too, were the studios which would be used—in this case Riverside 1 for all episodes except the second, which would be transmitted from Television Centre 1 (Cat. C)*—and the number of workshop man-hours allocated by the PBE for construction of the sets. The production as a whole was allowed 2,650 man-hours, rising subsequently to 2,800. With these hard facts before them, and after some generalized leads from his producer ('I want a big tree in the park'), Abbott began to sketch out tentative areas of action on squared-up studio ground plans. Almost automatically he would keep in mind such permanent factors as the availability of service points, access to fire exits, the positions of quick change rooms and the scenic transit dock. He would privately have noted such others as were peculiar to this production: the change of studio for Episode Two and the need to concentrate workshop effort if so many scenes were to be squeezed from the budget. Studio plans were torn up, sketched out afresh, amended, discarded and reconsidered. At the end much detailed designing remained but the acting areas had been established with sufficient finality to enable Abbott to hand over his notes and roughs to his assistant to draw out a set of working ground plans. After this meeting no radical changes could be entertained: these were the areas to which, during November, delineated by rows of chairs and lines of adhesive tape on the floor of a big drill hall in Wimbledon, the cast would learn to adapt their every movement.

The scripts, with single-spaced dialogue on the right-hand half of the page only, were almost bare of stage directions. Such as crept in were brief and emotive: 'The dust flies in the room', 'The sun is streaming in', or 'A dark room. Heavy with curtains. A heavy room. Pictures on the wall.' In fact

* By trade union agreement the amount of space to be utilized in a big studio such as Television Centre 1 is categorized; if for manpower reasons less than the total space (Category A) is to be employed, director and designer must not stray by a foot into the vacant space so tantalizingly ringing their sets.

there was no shortage of information, for Dostoevsky's book is charged with vivid detail—'it's all there' says Abbott—but the brief indications of the script are nonetheless close to the heart of the television designer's problem. Television, it has been said, is concerned only with people: more—that its only real demand is a human face in big close-up, two eyes and a mouth talking. A gross overstatement, it may be, but serving perhaps to put the designer's contribution in perspective. It is not his job to create the three-sided room of the illusionist theatre. Nor has he elbow-room for the great multitude of backgrounds demanded by the cinema, let alone the supercolossal sets of the wide screen. Rather must director and cast be provided with a limited number of fragments of backgrounds, of greater or lesser complexity, all within the confines of an area perhaps 60 or 70 ft. square, through which the actors can move in what is, notwithstanding film links and recording breaks, a continuous, consecutive performance. The effect of these fragmentary details is cumulative. Each in itself may be of the utmost simplicity yet, by a sensed authenticity, help to build up a powerful sense of place, of occasion, of atmosphere, as the story develops. To achieve this visual authenticity, great experience is called for on the part of the designer and his lighting technicians. His significant detail must be infinitely more refined and sophisticated than anything demanded by the theatre; yet, paradoxically, it may be contrived by a simplicity of means that would be impossible in theatre or cinema. Visual conviction may reside in the tangible grain of a wooden door or the glint of a metal surface as the camera tracks round it; it may equally be found in some roughly mounted, life-size photographic blow-ups of an actual interior. Nor must the effect of the sets upon the players be forgotten. Authenticity of atmosphere will assist authenticity of performance. For *The Materialists,* a play about Auschwitz, Abbott had had no compunction about filling the studio with mud. There are no rules in television design; only a need for true visual awareness.

In *Karamazov* Abbott had designed a series of units which could be assembled in different ways to form different sets or elements of sets. Concentration of resources upon a limited number of components not only enabled him to achieve a higher standard of finish than would otherwise have been possible; it formed, he felt, an elegant solution to the design problem and gave coherence and unity to the production which could not have been achieved in any other way. He determined upon the same approach this time.

Having developed his thoughts about possible elevational treatments, he had first to see what items might be available from stock, for television, with its voracious appetite, finds it worthwhile to maintain large quantities of material for continual re-use. Most of what existed in the Scenic Servicing Department's stock rooms at Alexandra Palace, however, was coarsely detailed and lacked the character and sense of quality which Abbott sought. He therefore next paid a visit to Pinewood film studios, outside London. Film studios' scenery is mostly custom-made, well detailed and expensive; from Pinewood he was able to select, on hire, a considerable number of items, among them 'one railway compartment, Spanish; one Russian stove; one bandstand top; four Gothic pillars'; and no fewer than twenty-one doors of various kinds. Armed thus with a knowledge of how much suitable material could be obtained and how much would have to be constructed, he returned to the inner sanctum of the third-floor room where he worked with his Group at Television Centre, to evolve the sets in greater detail. On the pin-up board, among the Work Variation Orders and the injunctions to mark correct zero hours on drawings, now appeared a breakdown of the man-hours available for each of the five episodes of *The Idiot.*

His aim was to make as much as possible of his scenery usable throughout the five episodes. To this end he evolved a series of nine units, of each of which two, three or four off were to be ordered. Seventeen other items would require construction; in all, some forty constructional drawings would have to be made. The bulk of all this would be needed for the very first episode for which one thousand man-hours of his total allocation was therefore allowed. This was to lead, a week or two later, to requests for supplementary allocations.

The immediate next stage, however, in November, was the first of the five vital Programme Planning Meetings called by the Director, Alan Bridges, to discuss the five-part production as a physical operation in the studio. To this meeting came, in addition to Bridges himself, his assistant and his secretary, Tony Abbott and his assistant designer, Robert Wright the Technical Operations Manager, his sound supervisor, the Costume Supervisor and the Floor Manager (who would take charge of the operation on the floor of the studio as directed from the Control Room). Though Abbott, Wright and the Sound Supervisor would all find themselves attending outside rehearsals to check on actors' groupings and movements, and to snatch a consultation from time to time with Bridges, this was the first and last occasion they would all meet together until they found themselves in Riverside 1 on 7 December. Alan Bridges sketched in quickly the general background to the production and indicated broadly how he intended to approach it. Abbott then spread out his drawings and described the nature of his sets and their disposition in the studio. On his gridded ground plan, Bridges indicated in pencil, using his plastic stencil with the cutout shapes of monitor, pedestal camera, powered dolly, piano and a score of other items, his tentative thoughts on camera positions and movements, the probable movements of actors, the most suitable positioning of microphone booms and so on. Each technician considered the problems arising in his own field, sought clarification of any obscurities, made copious notes, offered suggestions, requested amendments. This meeting was the catalytic point of the production. Thereafter, save only for any changes which the director might find it necessary to make in the studio, the framework of the production would remain unaltered. Every member of the team would be working independently on his own contribution, in full knowledge of how that contribution would relate to the operation as a whole. In reality, however, things were not quite as smooth as this. Already the timetable was slipping. Bridges was not entirely happy with the script and one complete episode was being rewritten. As a result, he was at one point no less than 27 days late with the essential information required by the rest of the team. From such situations do the extreme pressures in television arise.

The November meeting was also the first, as far as *The Idiot* was concerned, between Abbott and Wright. The men of BBC's 'Tech Ops' are regarded broadly as handling a series of mechanical, if complex, processes. They enter as trainees, are given a thorough grounding in electronics, and can then become cameramen, lighting men, boom operators and so on. They have infinite possibilities of promotion within the technical field but are not regarded as 'creative' within the studio team. Camera operators, for example, are laid on in the studio* much as water, power or any other service. Sensitivity is still required, however, in certain aspects of tech. ops and nowhere more so than in lighting. Lighting can make or break a designer's sets, and through them a whole production. In the last analysis the sets can be dispensed with but not the lighting.

Time was when those who took up lighting did so because they wanted to. Now, for some, it may be no more than an additional chore, without knowledge of which promotion up the tech. ops ladder would be impossible. In this respect, as others, Robert Wright was well matched to his Director and Designer. As 'TM I' he would be in charge of all tech. ops in the studio. In addition, however, he was not only himself a brilliant lighting man in the purely technical sense ('He came as a trainee', says the manager of his Department, 'and now he outshines us all') but was temperamentally attuned to the degree of improvisation they were apt to demand. Bridges and Abbott knew that whatever they might ask him to do, even at the last moment, he would do it, probably with his own two hands, willingly and brilliantly. Just such an occasion was to arise in Episode Four of *The Idiot*.

Immediately after the Programme Planning Meeting, such minor adjustments as were called for were made to the set designs, drawings were printed,† and a whole range of activities began simultaneously. Owing to pressure on the BBC's own workshops, the construction of the scenery was given to an outside firm of contractors who specialize in theatre and exhibition work.

* Outside Broadcast operators, and television's own film men, are regarded rather differently: they are the duffle-coated brigade who are expected to climb chimney stacks and use their initiative when they get to the top.

† Contrary to architectural practice, all information is wherever possible conveyed on a single drawing.

The Property Master had to be seen and arrangements made with the Buyer for special items. The green foolscap sheets of property requirements finally totalled nearly 650 different items, ranging from 'enough mashed potato for one vegetable dish at 3.00 p.m. and 7.30 p.m. on Wednesday, 8 December' to '6 live spiders; 2 very real-looking dummy spiders to be seen in c.u.'; from '144 sandbags' to '1 reel of black cotton'. 'Graphics' requirements were sent through (in fact, they were undemanding, consisting only of plain caption cards for the main titles and a roller caption for the credits). For the next month Abbott and his assistant were involved in a mass of detail, visiting workshops, selecting props, dashing off to rehearsals at Wimbledon, making out variation orders, eliminating items in order to keep costs down.

During these five weeks their lives took on a certain rhythm, building up each week-end to the culminating effort in the studio. Each Monday morning the scenery was delivered to the transit area of Riverside 1 or TC 1. By midnight the studio would be cleared of its previous show. The floor was empty. The great lights on their bars would be lowered from the roof, for replacement and re-rigging as required by Wright's plan. The main scenic units, numbered to accord with a marked plan, the grid references of which in turn accorded with similar references on the floor and walls of the studio, would be stacked in predetermined order. Properties and furniture would be brought in by big wheeled cages. In the hands of the setting crew the scenery would begin to take shape. Floor surfaces would be painted where necessary. Abbott would be here, there and everywhere, roughly positioning furniture and drapes, making a note of some lighting possibility he had just seen to put to Wright in the morning; listing retouching and other painting jobs that would have to be done next day. Mostly it was about 3.30 on Tuesday morning when he left to get some sleep. Sometimes the timetable would be reversed, as for Episode Three, when he worked through from 3 a.m. until 10 p.m. the following night.

Cleaners and maintenance men would come and go. At 9 a.m. the production crew arrived, followed by the electricians, painters and carpenters. Wright would begin to adjust the lamps for a first trial and tentatively set the 'memories' by which certain preset combinations of lights could be brought instantly into play. Abbott would now concentrate more sharply on the camera's view of the sets, placing himself systematically in every proposed camera position with a view to refining and adjusting where possible the image presented to that angle. Wardrobe staff appeared. Cameras, sound booms, control room links and circuits were tested. Actors arrived. By midday Alan Bridges, seated with his production staff before the banks of monitors in the glass-sided control room, high above the studio, in touch with the technicians and Floor Manager on the studio floor through the headphones they were wearing, was ready to start his first run-through.

The two days in the studio are regarded as camera rehearsals and this first run-through is always laborious. Its purpose is to line up each shot from the scripted camera position, stopping the action whenever necessary to clarify an instruction, to adjust a move so as to eliminate the shadow cast by a boom, to re-time an actor's pause, to ease a mix to another camera. Stage hands have duties to learn for the first time. The tone of a costume may be so close to that of a background as to require the latter to be lightened or darkened. The designer now has one eye all the time upon the studio monitors, in order to judge the image that will be transmitted. His greatest ally is the Lighting Supervisor, who alone can bleach out the junction between cyclorama and studio floor, alone can create the fragile actuality of a particular moment of time when the weather was such-and-such, the season so-and-so, the time of day thus. His greatest enemy is time.

Up in the control gallery, in three linked rooms, fifteen people sit, for the most part impassively, in total concentration before the pale flicker of the monitors. On one side the sound engineers are balancing, second by second, the volume and quality of what is coming through from the studio, cutting out one mike, bringing in another, feeding in a tape of a railway carriage travelling at the appropriate speed, mixing two particular bars of music, adding echo. On the other side, in vision control, the green, white and blue bulbs wink in the lighting grid, as the twenty memory combinations call into play the different kilowattage studio lights they represent and, beyond the glass, precise areas in the studio leap to life, fade and are transmogrified. On a bank of monitors the signals from the cameras are continuously balanced.

In the director's central control are seven more monitors: five taking signals from the cameras in the studio, one for film, one showing what is actually being transmitted or recorded. From the director himself flow instructions, warnings and requests, as conversationally as at a vicarage tea-party.

'Very nice, Dick. Crab right. Leave room on the right for another figure to come in. Stand by, Spencer. As soon as she's there. That's the right one. Come in, Two. Must be much tighter, all four together. Cue film. Stop that man walking about. Cue and punch up, Jill. It's midnight, Bob Wright, not midday. There's no light behind the window. Can we have a bit more echo on this, Richard. Cut. What happened, Alec?' Indecisive gefuffle. Then the Floor Manager's voice: 'We thought it was 35c the door.' 'Right. Let's go back to 33 and take it at speed. Cue film. Come in, One. . . .'

Bridges, Abbott and Wright were now in their element, using the components, human and inanimate, they had created, shaped and assembled, to realize in the studio the web of sounds and images they had first glimpsed in their discussions months before. In a sense the creative work was behind them; no happy accidents now could save the production if their work to date had been ineffective; just as no miracle of production engineering can transform any other product if earlier design development has been inadequate. On the other hand, much could happen in these twenty-four hours to deaden the tenuous atmosphere so painstakingly built up, to blur and blunt the essential communication at the heart of the story. At all costs the cutting edge of the production must be kept sharp and bright.

The first stop-start run-through is normally followed by a costume rehearsal without break, to test the production mechanics so far established, to test the continuity of the shots decided upon, to test the timing of the programme. Particular sequences may then need further rehearsal for further amendment or polishing but thereafter the programme is normally regarded as complete. Bridges, as we have seen, was a perfectionist. Work on each episode continued throughout every Tuesday and every Wednesday. Such was his own enthusiasm, however, that an air of exceptional excitement was engendered in his cast and his production team, so that even the camera operators responded to the situation with a special alertness and precision of framing. Up and down the narrow stair between studio and control gallery Abbott, Wright and he shuttled. Some directors at this stage work almost entirely from the control room. Bridges came down on to the floor a lot, to sort out snags at first hand. In all parts of the studio there was a sense of participating then and there in the act of creation, rather than mechanically acting out roles that had been pre-ordained at other times and places.

Yet, for some at least of those working at this high pressure, a part of their minds had still to be occupied with the problems of the weeks ahead. On the very day that the first episode of *The Idiot* went out, a memorandum came to Abbott telling him that 'It has been agreed that this serial shall be allocated an extra 150 man-hours, as under. . . . Under no circumstances will you be allowed to exceed the allocation for each episode on this serial. So you *must* design accordingly.'

'I want a big tree', Bridges had said. The way that Abbott gave it to him and Robert Wright lit it for him may surely be singled out from the rest of the production as providing an object lesson in sureness of touch and scale; in the multiple use of limited space to give flexibility of movement and a sense of infinitely greater space; in sheer integrity of effect.

Episode Three of *The Idiot*, 'The Trees of Pavlosk', called for two interior sets, an exterior veranda scene and garden walk, and a park with a bandstand. All these had to be accommodated in a studio 68 ft. by 64 ft.—say the length of a cricket pitch in both directions. Abbott gave half this space to the creation, in miniature, of a park-like exterior and used this as a natural backing to his interiors. More or less centrally in the 'exterior' half of the studio he constructed the trunk—12 ft. high and more than 4 ft. in diameter—and lower branches of a great cedar tree. Above this, 15 ft. from the floor, was hung a hexagonal frame constructed of 'drape bars', for attachment to which the Property Master produced twelve large fir branches. Around its foot with the aid of 144 sandbags was built a gentle mound; over this were spread 144 green turves and the contents of three sacks of dried leaves. To one side he constructed the octagonal bandstand; to the other, in false perspective to give a greater sense of distance, the garden path leading to the veranda of Lebediev's villa. Twelve smaller trees, lifted the day before; eighteen shrubs; a garden bench and eight statues on pedestals added their various notes to the composition. Facing on to this green oasis, in the other half of the studio, were constructed the two interiors of Epanchin's and Lebediev's villas. Linked to their grounds by verandas, french windows, conservatories and double doors, these rooms revealed altogether unusual depths of perspective and conveyed a powerful sense of being set in a real and mature landscape.

To have poured such a quart into the pint pot at his disposal was an achievement on Abbott's part. Without Wright's lighting, however, the result could have remained a makeshift thing. For Wright rules, and television has stern electronic rules, are made only to be broken. He is concerned to extend the limits of the medium, in particular the tonal range that is accepted as proper to television transmission. Outside his rooms the daylight is stronger than the light within, so that figures seen at a window against the sunlight beyond at times appeared as no more than silhouettes. Through the foliage of Abbott's great tree he poured sunlight, bleaching out the distant sky in a shimmer of heat and dappling the picnic party on the leaf-strewn grass with tremulous light and shade. At night, in the gloaming, as the girls in their long dresses and the tormented young men strolled past the glimmering statues, the trees of Pavlovsk took on an altogether different air of mystery and sadness. Pervading the whole production, and matching the integrity of the playing, was this same sharp sense of place, of actuality, of a particular evanescent moment of time. It was the heightened reality that results only from true sensibility. Christmas came and went. The Wednesdays slipped by, each week bringing its own problems, its own solutions, its own improvisations. In Episode Four, for which the design allocation was a mere 100 man-hours, Abbott had somehow to conjure up a street scene in ten minutes. Bob Wright lit it in two. The result was minimal; what would lend it instant authenticity? A dog, brought by taxi, arrived as transmission began.

And thus *The Idiot* ran its course. Nastasia lay naked and dead upon Rogojin's bed, her wedding dress billowing from a corner of the four-poster, the curtains at the window suffused with light from the living world outside. At 11 p.m. on Wednesday 12 January the cast removed their make-up, changed and gathered for a farewell drink. The production was in the can, ready for transfer to the radio waves which would bear its fragile imprint as they winged their way through space. The set was struck, the props wheeled out. At midnight Riverside 1 was clear and ready to receive its next production, already being shifted in from the transit dock. Another night's work began.

11. A Meeting of Minds

Such, then, is the kind of context in which design is carried out. In so wide a field no half-dozen examples could be other than arbitrary. We have nonetheless seen something of the way different teams of skills play their part in the design process: some of them large, some of them small; some embedded within big organizations, some in private practice; some under the same roof, some widely dispersed; some permanent, some assembled for a particular project only.

And so? The mounting of a great exhibition, the construction of a group of buildings or a great ship—is it not self-evident that complex enterprises such as these must call for the harnessing of many aptitudes? Almost every useful activity of civilized man, as well as many a useless but pleasurable one, is the result of group effort. Every commercial concern that is larger than a street market stall, every agency of government, every public utility, all public sport and entertainment, is the sum of the efforts of many people: sometimes hundreds of thousands of people grouped in vast industrial and administrative complexes. In what respect does the concept of group practice in design differ from these?

It is no part of my purpose to imply that virtue resides in size as such, that quality and size are necessarily related, that large teams must design better than small ones. There are wide areas, in surface design and in the craft-based industries more particularly, where individual sensibility will always remain supreme. Individuals, small offices, small teams will always produce the best fabrics, wallpapers, silverware, ceramics, posters and graphic design of every kind. They will produce the most personal furniture, interiors and stage sets, landscape schemes, buildings and structures. They will almost certainly be cheaper to employ, for their administration will be simpler and their overheads lower.

This is not really the point. What is really at stake—and this is as true of Western civilization generally as of the particular aspect under discussion—is an attitude of mind, what I have called our patterns of collaboration. Because some jobs are best done by the individual, or by two or three individuals in close relationship, does not mean that such jobs exist in isolation from a wider design context. One man designs a poster but that poster is part of a series; the series forms part of an entire graphic identity as evidenced by the organization's other advertising, its publications, the surface design of its goods, its delivery vans; these in turn cannot be unrelated to the image of the organization as reflected by its factories, its offices, its shops, its stations. Is it possible for these things truly to share a common design identity unless those concerned are brought into contact jointly to formulate that identity? The whole movement of society is towards the creation of larger units, including larger design teams, whether or not they are recognized as such. Certain types of design (pre-eminently environmental design) can only be done by big teams in concert for considerable periods, but even in much more minor spheres, wider groupings are being forced into being, if only temporarily for limited objectives.

What this book is about, then, is not big design offices as such. It is about communication and responsibility and creativity and collaboration. It is about ways in which the human scale of endeavour and concern can be retained in an increasingly inhuman scale of work. This is the central problem of our society and our time. Our objectives, and the machinery set up to gain them, can only get bigger, and more complex, and embrace wider and wider areas.

> The real question before us is whether this high degree of 'bigness', and the planning that must accompany it, call for the subordination of the individual to an external and uncontrolled bureaucracy, or whether democratic controls cannot be improved or newly devised for every level of our working and social lives that would permit the active participation of the citizen in deciding upon the course to be followed in those matters in which he himself, whether as producer or as consumer, is most directly and unmistakably involved.*

The solution is not easily found in the industrial, administrative or social contexts. Design poses further difficulties. For the most part the functions exercised in day-to-day life call for no particularly creative ability. The repetitive patterns of administrative and executive work, as of production processes, are clearly established; the participant may expect to meet a set of known requirements within a more or less restricted range. Original thought, on the other hand, is a lonely activity. The testing of a hypothesis is a logical process in which a group can participate straightforwardly. The setting up of the hypothesis is inspirational, the outcome of intuitive apprehensions stabilized by personal memory patterns. The creative moment, as we have seen, is that instant when, sometimes by a supreme effort, sometimes without warning, every factor affecting the total problem and capable of apprehension by the intelligence concerned, is held in the mind simultaneously and appears before the inner eye arrayed in a single, synoptic vision. This is essentially a private experience and is not easily to be shared out among a team. Herein lies the special character, the difficulty and the fascination, of collaboration in design.

Not that the problem is confined uniquely to design. There are fields, no less creative in their fashion, in which group effort is so familiar as to be

* Professor Max Beloff, in a letter to *The Times*, 10 January 1966.

taken for granted: for example, sport, music, the film and television studio, journalism. They range, it will be seen, between the poles of self-expression and highly sophisticated problem-solving. Common to them all, however, is the need for a common accord, a meeting of minds, within the group.

This is most apparent when the time scale is speeded up to the point where decisions have to be taken simultaneously. The footballer, though subject to the discipline of a well-understood set of rules and the authority of his captain, is faced, moment by moment, with a developing situation which results from the improvisatory action of twenty-two men, the forward details of which can only be foreseen to a very limited degree. The part each individual plays in this group activity is comparable with that of the players in free jazz. The thing that results, game or music, cannot be pre-planned; it is created co-operatively by an accumulation of personal decisions taken by all concerned up to that point. Random action is prevented not so much by the power of the captain, or leader, who, despite his strategic and tactical orders, can foresee the course of future events with no greater clarity than his men, but by the limitations on the area of personal decision accepted by all those involved: tempo, key, etc., in the one case; the rules governing offside, handling of the ball and so on in the other. This is probably as close as modern man can get to the unquestioned aesthetic framework within which the archaic sculptor or medieval stonemason worked. The problem-solving element is missing, however, and this makes direct comparison with the work of the designer unprofitable.

The theatrical arts, as we have seen, call for both arduous pre-planning and the quick reflexes of spontaneity. They can range from complete improvisation (*pace* Cassavetes and *cinéma vérité,* a potential reality only in the live performance*) through the music-hall, ballet† and the drama to the multi-modified product that is the normal film.

The shortcomings of the traditional West End production are notorious. The team assembled have scarcely begun to know each other, let alone develop a common style of playing, before they are disbanded at the end of the play's run. Against this can be set the deep *rapport* of famous companies, such as the Moscow Arts Theatre, which have developed their repertories over a long period in conditions of stability and continuity. In such collaboratives as those of Miss Joan Littlewood or America's Living Theatre the whole company contributes to the development and production of the original script-concept during the rehearsal period and subsequently.

The common bond of empathy which unites such a team—or indeed that

* Though it is salutary to remember that only exceptionally was a scene in a Mack Sennett comedy re-shot a second time.

† An odd experiment was recently conducted by the Ballet Rambert, in which none of those concerned with the creation of a new one-act ballet—designer, composer or choreographer—had any personal communication at any stage.

which produced *The Idiot* for BBC Television—is essentially emotional. It derives its motive power from the heightened nervous tension engendered by that proximity of disaster which can never be far distant from any live performance. As soon as the time scale is slowed, what de Chirico has called 'the metaphysical assurance of the group' is slackened and the members of the team tend to become individuals again. Most design, as we have seen, can be a very slow business. A more relevant comparison is perhaps therefore to be found in the editorial office of a newspaper or periodical.

In any journal, overall editorial direction must clearly be established if self-contradictory contributions are to be eliminated, either by *diktat* and the staff officers of the sub-editing hierarchy, or by prior agreement voluntarily reached in joint conference. It is not hard to recognize those newspapers in which certain rigid policy lines, often rejected by staff reporters and writers personally, are imposed from above with the utmost inflexibility. Such offices form the counterpart to the architectural practice of the *prima donna* designer, supported only by draughtsmen and technicians. The pattern they exemplify is not, however, an inevitable one. Journalism on *The Times,* Sir William Haley has said,* 'is a corporate not an individual effort. So far as the views of the paper are concerned, they are never the views entirely of one man, they are the views of a number of people who form a group of which they think of themselves as members . . . there's a continuous process of influencing and being influenced in one's ideas the whole time.' *Picture Post* in its heyday may be instanced as a quite different type of journal whose policy was reached, for the most part, by common accord of the editorial staff. Its weekly conference was attended by editors, writers, photographers, layout men all along the line to the head of the typing pool and the office boys. Everyone was free—indeed expected—to contribute criticism of the issue that had just gone to press, ideas for the next and the future, and comment on all matters arising.

The elements comprising a particular issue were thus liable to development and modification by group action at a number of points: first by discussion at the weekly conference; secondly at the more detailed briefing of writer and photographer before they went out on the story; thirdly by agreement between the latter while *on* the story, in the light of circumstances as found (it was not uncommon to encounter difficulties in establishing a *picture* sequence that properly embodied the abstractions of an otherwise acceptable *word* story); fourthly by editorial shaping of words and pictures (often several quite different, even contradictory, stories could have been distilled from the many hundreds of photographs turned in by the cameraman: their selection and integration was the responsibility primarily of the art editor); fifthly by the physical requirements of production and by the editorial requirements of a specific issue (for example, a right-hand page opening to a feature would

* In an interview with Kenneth Harris. *Observer,* 8 May 1966.

differ from a double-spread opening, the 'tempo' of a story would have to be considered in relation to that of the features before and after it, length and balance might call for drastic reshaping in the light of last-minute inclusions in the issue). A minimum of six people were thus inevitably involved directly in the final shape of the simplest *Picture Post* story—editor, assistant editor, art editor, photographic editor, writer and photographer—and more often than not considerably more. Differences of opinion, if not settled at lower levels, would finally reach the editor for decision. Real disagreement, however, was rare. Accord was normally reached by a meeting of minds, born of an exceptional team spirit and, to all intents and purposes, continuous communication between those concerned with a story from the moment of its inception to the moment it was 'put to bed'.

The list could be lengthened. Group purpose is increasingly recognized, for example, by the award of Nobel Prizes to research teams, it being more and more difficult to distinguish the precise limits of the contribution made by each individual member of the team. Common to all such groups, in whatever field they may practise, is the degree of empathy that exists between their members. The acceptance of a group aim supersedes any purely personal, sectarian or single-discipline objective; by the same token the area of responsibility is enlarged from the personal or single-disciplinary to embrace the total operation. Common standards come into being and members of the team are forced continually to project themselves in imagination into the positions of their colleagues, and to see the operation from the latter's standpoint. Amongst the members of a multi-disciplinary team, Richard Neutra has written,* a wink of the eye is sufficient to seal a common understanding. To achieve such a harmony, however, three broad requisites are called for. They are ease and precision of communication; a clearly understood demarcation of responsibilities and administrative structure; humility and a readiness to accept a degree of anonymity on the part of the team members.

The problem of communication has already been touched on. It has been suggested that there exists today, embodied in the new tradition of the modern movement, the essentials of a common language for use by architects and designers—save only perhaps those concerned with graphic and surface design. As is only to be expected in an age of rapid transition, this language is still developing, being based not upon a detailed canon but upon a handful of principles. If, however, its loose and anti-Platonic approaches do provide a framework within which three-dimensional designers generally can communicate with each other, its rule does not extend as yet to engineers on the one hand, to painters and sculptors on the other. Chagall's ceiling for the Paris Opera has been likened to 'a renovated bistro'.†

* 'The Creative Team', *Architecture and Building*, June 1956.

† *Architects' Journal*, 30 December 1964.

Basic terminology differs between the professions; the very words hold different shades of meaning. To an architect 'design' means, first and foremost, the imaginative manipulation of solids and voids so as to create an environment. To an engineer, it generally means an analytical process, the aim of which is merely to confirm or otherwise the propositions put to him. To a sculptor it means the degree to which his piece successfully expresses a proposition of purely personal significance. None of the three, in all probability, has been educated to appreciate the nature of the thought processes of the others; none was trained to work as a member of a multi-discipline team; none was taught to see beyond the sub-problem with which he is primarily concerned to the total problem of which it is part.

This is precisely what present attempts to establish common curricula in training are seeking to break down. That there is nothing impossible about such an aim in principle is shown by that handful of constructional engineers —for example Felix Samuely and Ove Arup in Britain—who have shown an instinctive aptitude for working with architects and designers. Possibly more significant, just because it is more unexpected, has been the collaboration between the painter Victor Pasmore and the architects of the New Town of Peterlee. This 'act of faith' by Peterlee's General Manager arose from a realization of the shortcomings, in an environmental sense, of the town's first housing areas, already then embracing over 2,500 houses. His instructions to Pasmore and, in the first instance architects Peter Daniel and Frank Dixon, were in effect: 'Do what you like but don't do what we have done before.'

This first collaboration (Daniel and Dixon have since moved on though Pasmore remains as Consulting Architectural Adviser) began in September 1955 and lasted for about five years. Pasmore has stated that no idealistic conclusions should be drawn from it. Nonetheless there was between the three, it would seem, 'a complete synthesis in mind' and Peter Daniel has since said: 'The essential success of our collaboration was that all of us were prepared to work together and to a certain extent to submerge our individual identity to achieve a common end.'

The team were concerned with the south-west area of the town, nearest to Castle Eden Dene, a deep wooded gorge running in from the coast. Here they sought to develop a housing community of a robust and unsuburban kind, in sympathy with the landscape and more close-knit than had then been attempted in other British new towns. Only after an independent report on their proposals had proved favourable was Ministry sanction to proceed received. The end results were not without their faults. A number of practical shortcomings became evident and, in Pasmore's own words, the detailing was 'tragic'. Nonetheless, this marked the first attempt in the British new towns to establish an alternative to the garden city idiom and, as such, held considerable significance. Was this freshness due to the inclusion of an artist in the team? Pasmore says that he has done nothing at Peterlee which an architect could not have done. In truth, the degree of integration in the team seems to have been such that we shall never know. 'We did not divide into our specialized subjects', said Peter Daniel; and, as the various design stages were worked up, 'each personality', said Frank Dixon, 'had to a certain extent to immerse himself in the problem'. Pasmore's prior concern with the mathematics of ordered relationships in two and three dimensions undoubtedly fitted him peculiarly to work with architects in this way. The point remains that there seems to have been no problem of *communication*.

The second prerequisite to harmony in group working is clarity of administrative structure and demarcation of responsibilities. A three-month study on communications in the building industry, undertaken in late 1962 and early 1963 by Dr Gurth Higgin and Neil Jessup,* found that a principal cause of the present confusion in communications lies, not in the means employed (though these could be improved) but from a general uncertainty, shared by all participants, about their relationships and competence. This is not a problem that can be solved at the design stage alone. If administrative and creative pipelines are not to fur up, doubts and confusions, absorbing time, energy and enthusiasm, must be flushed from the whole system all the way from Board to chargehand. The primary conditions which alone make possible the creation of good design can only be established by management and effective links must be forged to bring management into the creative process.

Though the LCC record shows that it need not be so, too often design consciousness at Board level still springs from the chance presence of one exceptional individual: from the passionate concerns of a Frank Pick, an Olivetti or a Colin Anderson. By a curious twist, the senior partners of a practice that has grown to the size of SOM—tycoons in their own right and able to meet client tycoons on an equal footing—can sometimes take over the impresario role. Time and again SOM have enlarged the horizons of their clients and created a desire for something better than was ever envisaged at the outset ('Don't be cheap', says Merrill; 'Make no little plans', says Bunshaft). It is relevant, in this connection, to recall Dr Maldonado's plea that we should put aside

> the prejudice that only one kind of architect can exist: the kind who designs buildings. We should train other kinds too: architects trained to study sociological and economic conditions of housing, not as sociologists or economists but precisely as architects; architects with enough knowledge to be fused into the planning, standardization and dimensional co-ordination offices; architects who will be able to take over managerial positions in the building industry, and who could from these vantage

* Of the Tavistock Institute of Human Relations, at the request of the NJCC. Tavistock Publications, 1965.

points watch over the cultural interests of the end product; the human habitat.

In Britain, save in the field of civil engineering, such penetration into management is prohibited by the professional codes of the institutes concerned.* Nonetheless, in industrial design no less than in architecture, professional qualifications and experience are sorely needed in design management.

That steering and advisory committees when thus constituted can prove enlightened allies, bringing into sharper focus issues that might otherwise remain clouded, is evident from the cases already described. In the Festival of Britain, in working on *Oriana*, in establishing the requirements for the Chelsea School of Art, the design teams speak with admiration and gratitude of their committees. 'The one astonishing job . . . it really could not have been better', says DRU of the Orient Line Steering Committee. 'Inspiring', says Michael Powell of the Chelsea Advisory Committee, adding that its members had a full grasp of the problem and ability to formulate the brief—but were then happy to leave the design proper to the architectural team.

If it be vital for good communications and administrative clarity to bring management and client into the design process at one end, it is no less so to bring in the manufacturer or construction company at the other. The collaboration between Henry Dreyfuss and the engineers of John Deere and Company has been instanced in the field of industrial production. Building Design Partnership can provide a comparable example from the building industry. It has long been this firm's policy to integrate with their contractors as early in the programme as possible. In an urgent industrialized building project, performance specifications were given to selected contractors who were invited to submit quotations on that basis. It was thus possible to select a contractor before beginning to design the project and his expertise was at the architects' disposal throughout the design process. This worked so well that when, subsequently, BDP decided to use an industrialized system of building for sections of Surrey University, 'members of the contractors' staff moved into the architects' office, and the traditional master-servant relationship between architect and contractor was completely abandoned'.†

Two factors may be noted from all the teams studied in these pages: one that in each case the administrative side of the organization was subsidiary to the design side; the other that the optimum size of the design unit seems to lie somewhere between seven and twelve, with the emphasis at the lower end of the scale. The former principle was exemplified very clearly in the Architect's Department of the LCC, a large bureaucracy where things might easily have been otherwise. The latter accords broadly with the findings of

* So that the Managing Director of SPAN, the first British firm to set a real standard in private development housing, had to resign from the RIBA to do so.

† From an article by Basil Honikman, BArch, ARIBA, in *Official Architecture and Planning*, September 1965.

Professor Robert Freed Bales, Director of the Laboratory of Human Relations at Harvard, who has given consideration to the structure and running of committees and has suggested that seven is the most satisfactory number.

Within the group, the more precisely a member's responsibilities are defined, the more precisely he knows what demands to make of his colleagues. In a small team, functions may be so sharply defined that responsibilities follow almost automatically. This was the case, for example, in *Picture Post.* If the operation is complex, with a whole series of teams involved in parallel, then some form of pyramidal control may be necessary. Dame Evelyn Sharp, now Baroness Sharp of Hornsey, has described the Civil Service technique as follows.*

> In any such group (of administrative and professional officers, research officers, engineers and so forth) you come to some questions where the different interests clash and the group cannot resolve them. When this happens the question goes higher to a headquarters group, perhaps at my level (Permanent Secretary) or 'Dep. Sec.' level, again with the other Departments, and with the senior professional officers.
>
> Differences are referred to a higher group, which is also a standing group . . . to resolve differences of opinion or disputes. And finally you come to a group of Ministers, which either may be a standing group of Ministers convened for the purpose . . . or it could be an *ad hoc* group to resolve particular problems which it has not been possible to resolve at a lower level. There is, as you know, a standing committee system of the Cabinet . . . and in the last and final resort, the arbiter is the Cabinet.

Group practice must not be thought to dissolve the need for leadership. Black, Dreyfuss and Rosselli, from the offices already described, are explicit about this. It is equally implicit in the other cases examined. Michael Powell had to throw his weight around in order to ensure that the Chelsea redevelopment came out in the unified way it did. The implications of leadership within the group context are examined more fully in the next chapter. In the meantime one senses here the third prerequisite to working harmony: a measure of humility among the members of the team and a willing acceptance of a degree of anonymity. The American sculptor, David Smith, not long before his death, wrote: 'Architects have the opinion that they are the fathers of all the arts, that their buildings are sculptures and that the use of paintings and sculptures oftentimes defiles their purities. There are no affinities between us . . . A man works better when he is working with his own spirit than when he is working with the domination or collaboration of the architect. . . . Mostly architects look down upon us and mostly they are

* After a paper 'Reflections on Planning' delivered to the Town Planning Institute and printed in the *Journal* of the Institute, June 1964. It should be remembered, however, that in the Civil Service, unlike the LCC, the administrative officer takes precedence over and is served by the professional.

big business men, and we're just one of their small clients in the building. They choose to put the marble in the men's room and the bronze in the fixtures and they don't really need sculpture at all.'* That this remains true of very many architects does not disguise the voice of the specialist without desire to see beyond his own sub-problem. It is the voice of despair.

It must not be thought that willingness to submerge self in reaching for a team objective necessarily leads to passivity.

> You have to come to the office, [says Denys Lasdun], and hear the infighting, the differences of opinion, experience the rhythm of design, see how the discipline of architecture has the miraculous power of reconciling differences between young people, middle people and old people. We are not a happy little band who go round agreeing about every problem we are working on. We argue, discuss and absorb information from many different skills. I bend over backwards to let the younger members of the office have their say and . . . when necessary prove me and my colleagues wrong. What does it matter who did what, providing the building is responding to its own system? . . . It is also a matter of time and working together and the thing that keeps the office together is that the balance of sympathy outweighs the balance of antipathy; but there must be both. [Those who stay] subscribe to a principle, an attitude, a process, because without these there can be no intellectual foundation for a consistent development over a whole corpus of work.

If this can be said of so personal a practice, how much more must it be true of the multi-disciplinary group—or at least the group in good creative health.

Professor C. A. Mace, Emeritus Professor of Psychology at London University, has described the essence of group practice and consortia as 'a diversity of talent and a unity of values'. 'When the professional frontiers can be broken down to the new situation (group practice) and a common identity of purpose established', says Philip Dowson of Arup Associates, 'there is a tremendous release of enthusiasm, particularly among those who traditionally have the less interesting jobs to do.' Sarah Harkness, a founder partner of The Architects' Collaborative, has written:†

> Architects who practise alone have doubts sometimes about a collaborative office, fearing that if they were part of such a group they would lose their individuality. In a collaborative office, however, something is added, rather than subtracted. Creative talents are stimulated by an 'active environment'. I do not think that the success of a collaborative office is dependent on a remarkable combination of particular people. It is a practical system—on the way to becoming more usual. If there is anything magic in the whole set-up, I believe a good deal of it may simply be a matter of numbers. The partnership is large enough so that it would

* Quoted in *The Architect and Building News*, 5 August 1965.

† *Architectural Review*, May 1957.

> be almost impossible for one person to dominate it. It is also large enough so that whatever idea one member may have, he will always find a willing ear to listen. . . . However, the group is small enough so that the partnership can meet round the conference table for informal discussions; small enough so that decisions do not have to be made by vote, but only by 'the sense of the meeting'. Our meetings have been likened to Quaker meetings in this respect.

Perhaps Matt Busby's* description of team spirit is as good as any: 'This is hard to define. I think of it as all working to help one another, sharing the work, having a sensible respect for every individual's capacity and the team capacity and yet to express your own individuality at the right time. Ability alone won't get you through.' Among the examples studied in the previous section, the production team for BBC Television's *The Idiot*—admittedly a relatively small and close-knit group—provides clear evidence of just such a *rapport* between its members, and what may be expected to flow from it. The construction of Islamabad, admittedly a much more complex operation by a much more dispersed grouping of talent, shows some of the dangers that exist if such an instinctive *rapport* is lacking.

If this subtle harmony of minds can be achieved, what results may be expected to flow from it? Clearly, group practice in itself provides no guarantees of brilliance. Quality must always depend upon the talent of the designers concerned; there is no substitute for the creative faculty. In its detail, the Festival of Britain suffered from gimmicky affectations which were to bedevil interior design for a decade. It cannot be pretended that *all* the public areas of *Oriana* welded into a perfect unity, let alone that the interiors were conceived as part and parcel of the same design problem as that posed by the form and structure of the ship herself. Michael Powell's problems in keeping his four-in-hand in step were such that, in the next comparable job for which he was responsible (indeed, one costing twice as much), he felt it necessary to jettison the multi-group structure and substitute for it a single, enlarged team.

A single, all-powerful designer, it may be argued, could have obviated such weaknesses. With equal force it could be said that such shortcomings might often have been overcome if the process of group collaboration had only been carried further. In the case of *Oriana*, for example, DRU themselves felt in retrospect that the individual designers working on different rooms—either within or outside DRU—were insufficiently aware of what the others were doing and that, on another occasion, it would be better to arrange regular meetings at which all schemes could be seen by all those responsible, if only as a psychological incentive. Further, it was a curious feature of this

* Club Manager of Manchester United, in an interview with Bob Ferrier, *The Observer*, 15 September 1963.

operation that the cabin design, legitimately handled by one design office, was outside the co-ordinating machinery set up for the rest of the public areas of the ship. Might not greater contact with the Steering Committee and the other designers have provided additional stimulus?

Again, as far as the total vessel was concerned, it was made clear that any overstepping of the traditional boundaries between interior design and exterior design would not be welcome. Ships come into being as part of a process almost like Hoyle's theory of continuous creation. There is no moment in time when a shipping company tables a set of requirements with precision and briskly decides to buy a new ship to meet them. Rather are their thoughts always turning to 'the next ship', which may be one, or five, or twenty years away. In the light of their collective experience and the changing conditions of the trade, this image gradually takes shape until the pressure of the concept is such as to be withstood no longer. Lo and behold, there is the steelwork rising on the slipway; arrangements can be made for a launching; and perhaps for a scrapping.

In this long development of provisional decisions and tentative assumptions it is perhaps inevitable that the interior designer is not called in until the nature of the boat has reached near-finality. It is interesting, however, to speculate, in the light of Professor Black's findings in relation to the Beagle aircraft, what might be the effect of building-in the designer at an earlier stage. It seems possible that new inter-deck relationships might be achieved internally; new forms of access, flow and circulation; certainly new forms of superstructure. The very framework of the ship could be evolved with certain spatial and decorative end results in mind, instead of the decorative concepts being wholly subordinated to the purely functional space-frame evolved by naval architect and shipbuilder.

Group practice, then, like the new tradition of design itself, is a method rather than a promise of perfection. Like the new tradition, it must be judged as much by its potential as by its performance. Like any other method, it must depend finally upon the calibre of those who use it. As a method, it has undeniable advantages, chief among them being flexibility in the short term, stability in the long.

It liberates such creative abilities as may exist at all levels throughout the office. The RIBA report on *The Architect and His Office* found greater *efficiency* to exist in offices organized as 'vertical' or hierarchical structures but that 'horizontal' groups tend to produce more freedom for the staff and work of greater originality. Group practice allows the leadership of the team to shift according to the skills of the team and as may be appropriate to the demands of the job in question; different leaders may even head the team at different stages of the same job, as was found desirable by Professor Robert Freed Bales in his studies of the small group at the Harvard Laboratory for Human Relations, of which he is now Director.

By the same token, expansion and contraction of staffs, shifts in emphasis from one field of design to another, can be encompassed within the office's normal pattern of working. When DRU brought additional offices into the *Oriana* commission, they were but enlarging the group—the administrative structure remained the same. Group practice not only brings in all concerned at the outset but permits maximum and continuous communication between them thereafter. By this multiple feed-back system, it constitutes an error-correcting device that will tend to check gross errors of judgement and so produce work of relatively even quality.

Most striking of all, perhaps, is the group's potential for long-term survival, with all the possibilities of building up a consolidated aesthetic thereby implied. Hierarchical offices eventually ossify or collapse. Consider the pyramid form for a moment as inverted, with a great weight of substructure turned superstructure bearing down upon the point of the single individual who leads the organization. Its inherent instability becomes immediately apparent. Remove the individual, let his powers of concentration and invention flag but momentarily and the whole balancing act crashes in ruins. Normal professional partnerships come and go. Architects in particular, it would seem, throw themselves into professional marriage and divorce with the indefatigable energy of Hollywood stars. The group, on the other hand, to which individual ambitions are subordinated, holds that inbuilt capacity for change and development which can alone ensure long-term continuity of existence. The group alone can develop that inner life of its own which persists through the sloughing off of some cells, the creation of new ones. The LCC no longer exists, but at its death the vitality of the Architect's Department remained unimpared in spite of two decades' exceptional turnover of staff. DRU, as an organism, pulses as healthily as ever it did twenty years ago.

The group's potential for continuity and collective survival is pinpointed by Skidmore, Owings and Merrill's partnership formula. This was established in 1949 when Skidmore and Owings took in Merrill, Severinghaus, Brown, Cutler and Bunshaft as general partners. 'It provided that no outsider should be permitted to buy into the firm. Partnership in SOM was to be restricted to architects elected unanimously by the partners from within the organization: specifically, from the two management echelons of "associate partners" and "participating associates" that were then set up under the general partners. To enforce this principle, the partnership articles deny any partner the right to sell his shares to anyone who is not a partner, and both the purchase price and the terms of payment must be agreeable to all the partners. (To make it easier for the new partners to buy their shares, a system was worked out allowing them to pay for their partnership out of future earnings, after allowance is made for living expenses and taxes.) If a partner retires or dies, furthermore, his partnership interest may be acquired only by the remaining partners or by newly elected partners.'*

The rationale of group working has much to commend it. It is the ideal rather than the rationale which draws men to this form of collaboration, however. Without pushing political parallels too far, one is forced to the view that the egalitarian rejection of an authoritarian chain of command is at base idealistic. No less so is the belief that only through identification of self with a purpose voluntarily shared can mankind move forward into the mysterious future with dignity and purpose. Man moves on the backs of his predecessors and his own contemporaries. Paradoxically, that the group provides the only rational way forward remains an act of faith.

Walter Gropius, as we have seen, was the first in recent times explicitly to give expression to this belief. Half a century later and half a world away, his faith in that vision remains firm. Here are some words of his, on the occasion of the conferment upon him by Columbia University, in March 1961, of the Degree of Humane Letters:

> A relatedness of expression and a consolidation of trends cannot be consciously organized in a democracy, but springs from spontaneous group consciousness, from collective intuition which brings our pragmatic requests and our spiritual desires into interplay. . . . But . . . the nineteenth-century idea that individual genius can only work in splendid isolation . . . the conception of the architect as a self-sufficient, independent operator, who, with the help of a good staff and competent engineers, can solve any problem, and keep his artistic integrity intact . . . is an isolationist attitude which will be unable to stem the tide of uncontrolled disorder engulfing our living spaces. It runs counter to the concept of Total Architecture which is concerned with the whole of our environmental development and demands collaboration on the broadest basis. . . . It is obvious that we have to learn new and better ways of collaboration. In my experience these call first of all for an unprejudiced state of mind and for the firm belief that common thought and action is a precondition for cultural growth. Starting on this basis, we must strive to acquire the methods, the vocabulary, the habits of collaboration with which most architects are unfamiliar. This is not easy to accomplish. It is one thing to condition an individual for co-operation by making him conform; it is another, altogether, to make him keep his identity within a group of equals while he is trying to find common ground with them. It is imperative, though, that we develop such a technique of collaboration to a high degree of refinement since it is our guarantee for the protection of the individual against becoming a mere number and, at the same time, for the development of related expression rather than of pretentious individualism.

There can be no doubt, of course, that the creative spark originates

* 'Skid's Row', *Fortune*, January 1958.

always with the individual, but while he works in close co-operation with others and is exposed to their stimulating and challenging critique, his own work matures more rapidly and never loses touch with the broader aspects which unite a team in a common effort.

Communication from person to person is at an all-time low today in spite of, or because of, our tremendous technical means of communication, and most individuals are driven into shallow superficiality in all their relations with other people, including their own friends. But just as the airplane is no substitute for our legs, so personal contact between people of like interests cannot be replaced by the vast output of professional literature and information service because individual interpretation and exchange is still essential for our functioning as human beings. Our over-extended receptive faculties need a respite so that greater concentration and intensification can take place, and I feel that a well-balanced team can help achieve just that. As we cannot inform ourselves simultaneously in all directions, a member of a team benefits from the different interests and attitudes of the other members during their collaborative meetings. The technical, social, and economic data, gathered individually and then presented to the others, reaches them already humanized by personal interpretation, and since all members of a team are apt to add their own different reactions, the new information is more easily seen in its proper perspective and its potential value. For the effectiveness of this kind of intimate teamwork, two pre-conditions are paramount: voluntariness, based on mutual respect and liking, and exercise of individual leadership and responsibility within the group. Without the first, collaboration is mere expediency; without the last it loses artistic integrity. To safeguard design-coherence and impact, the right of making final decisions must therefore be left to the one member who happens to be in charge of a specific job, even though he has previously received support and criticism from other members.

Such principles of teamwork are easier explained than carried into practice because we all still arrive on the scene with our old habits of trying to beat the other fellow to it. But I believe that a group of architects willing to give collaboration a chance will be rewarded by seeing their effectiveness strengthened and their influence on public opinion broadened. All teams so organized, I trust, will eventually act as ferments in our drive for cultural integration.

And again, also in 1961:*

A true team between architect, painter and sculptor, aiming at an organic synthesis of their work by symbolic association, needs an intensive exchange by its members. For only a thorough, mutual penetration into the ideas of the various team-mates may lead to that kinship of spirit

* Bulletin No. 41 of the International Association of Plastic Arts, UNESCO, Paris.

which can make a creative entity out of the individual contributions. The architecture and sculpture of the portal of Notre-Dame, supporting each other mutually, became one inseparable entity because, for the sake of God, the work had grown above its creators once they had lived and toiled together on the site to find a common denominator for expressing the sublime in human measure.

Perhaps, in the same degree as the idea of true democracy may more and more become a spiritual power and inspire the artist to give it representation, will a genuine collaboration between architect, sculptor and painter arise again. With a growing challenge there will also come an increasing clarification of the right training needed in the mechanics of integration and collaboration, and equally towards that desirable basic attitude that puts the whole above its parts, the ensemble before the individual. Then we may also have learned to comprehend that the present ignorant procedure, merely to commission painters and sculptors to adorn a building only after this has already been completed, was bound to prevent the greater potential results of true collaboration. If contributions of sculptors and painters are desired by a client, he should let the architect choose congenial collaborators at the initial designing stage of a building. Only such foresight will enable the architect to build up his team in time to let each member take part in the creative process of the basic space-conception, a process so essential to the result of final unity in the building.

The contradictions in man's nature are not to be satisfied by uniformity. The creative spirit is not to be confined by any single administrative formula. The British Motor Corporation revolves, managerially, around one imperial figure; the design of its cars stems from another. Conversely Ford, the second biggest motor manufacturer in Britain, operates on a whole-hearted team basis. The BMC's Austin 1800, according to Issigonis, 'was never discussed at boardroom level. Whoever heard of a committee writing a best seller?' The development work was done by one of BMC's three 'cells' (for small, medium and large cars) and the design of the bodywork by Pininfarina; both were under the direction of Issigonis, himself responsible only to his production chief, George Harriman. At Ford's, on the other hand, some fifty people are said to have been significantly involved in designing the Cortina. Their eleven-man Policy Committee embraces salesmen and stylists, engineers, production men and accountants. Intensive market research (more than the rest of the British car industry put together) precedes the design of any new car and 200 technical cost estimators work alongside the engineers from the beginning. The Austin 1800 took about two and a half years to create; the Cortina went through on its pre-planned schedule of nineteen months. Two cars reached by totally different routes: who is to say that one is better than the other?*

* I am indebted for some of the facts in this and the subsequent paragraph to two features

There will always be architects whose overriding purpose is the design of relatively few buildings of exceptional and personal quality. They may well restrict the size of their staffs to the point which enables them personally to supervise every detail of design. From such men may be expected the great imaginative leap, the breakthrough which makes all previous rule books out of date. A Ronchamp, Buckminster Fuller's geodesic domes, Moulton's grasp of the consequences of reducing the standard 26-inch bicycle wheel to 16-inch, Issigonis' transverse placing of the engine under the bonnet—such as these can only, perhaps, be the product of a single mind. 'Most people in the car business', Issigonis is quoted as saying, 'are out to make their models a bit cheaper or improve them in small details. I'm not here for that: anybody can do it. I'm here to make dramatic improvements.'

It is not, however, at this level that the work of the world is done. It is not at this level that our daily needs have to be met, that the physical fabric of our lives collapses into ugliness and squalor. We are too much preoccupied with masterpieces, with diamonds in the rubbish heap of our urban environment. Above all else our civilization needs to order the simple, repetitive elements of everyday life into a harmonious framework of considered relationships; a framework which does not bruise the spirit of man but enables him to live, play and move about his business with dignity and grace. Within such a framework the works of art and the endearing one-off quirks—the great buildings, the sculptures, the glimpses and vistas—will sparkle the more brightly to enrich our lives.

Let us not mislead ourselves into believing that genius is necessary to create an acceptable traffic bollard or a decent kitchen table or an avenue of trees. Genius is, by definition, exceptional and works of genius are the exception. Can we but recognize that the greater number of the design problems of the human environment ask of us no more than humility, co-operation and a spirit of anonymity, then will our surroundings reflect those basic good manners which alone can provide an appropriate setting for a civilized society. Then, and only then, will future generations, looking down the perspective of time, perhaps see in what we have made that same unity which we see in the great periods of the past; and perhaps, who knows, find here and there, neither more nor less than in the past, works of genius to uplift their hearts.

in the *Sunday Times* and *Observer* Colour Supplements of 25 October 1964; the former by Peter Wilsher, the latter by Graham Turner.

12. Leadership and Rewards: I

A note on Building Design Partnership

Building Design Partnership is the most resolutely multi-disciplinary team operating—in its field—in Great Britain in the mid-sixties. A glance at the organizational problems it has encountered, and at some of the solutions it has found to work well, may therefore be of interest. As befits a growing firm with its particular ideological concerns, BDP is at pains to avoid finalities in its philosophy or rigidity in its organizational structure. 'The strength of the firm', says David Rock, a London partner, 'is that nothing is finite or crystallized.' The details which follow apply, therefore, only to a particular moment of time. By the same token they must not be taken to have universal applicability. Their very flexibility, however, may commend the approach they embody to groups in other fields of design.

Between 1956 and 1959 the architectural firm of Grenfell Baines and Hargreaves opened offices in Manchester, Derby, Blackpool and London. In 1961 the idea of a comprehensive design group which had been forming over the years was formalized by the incorporation of a firm of quantity surveyors and a change of name to Building Design Partnership. Further offices were opened in Aldershot and Leeds and, in 1964, the seven then existing offices were formed into three groups—Preston Group (two offices), Manchester Group (Manchester, Derby and Leeds offices) and London Group (London and Aldershot offices)—to be joined in 1965 by the opening of an office in Belfast. In 1958 the number in the firm was 36; in 1966 it was 260, embracing planners; architects; quantity surveyors; civil, structural, mechanical and electrical engineers; landscape architects; furniture and graphic designers. Their disposition among the main Groups was, roughly, 85 in Preston, 35 in Manchester and 115 in London, with 25 in a Central Administration Unit in Preston which dealt with accounts and some bulk buying.

Establishing the right internal balance between the professions has raised points of principle and practice: horizontally, in the optimum numbers of each within a group; vertically, in the partnership structure and the need for each profession to be adequately represented at partner level. The original group structure, in the firm's first years, was as follows:

Architects	Quantity Surveyors	Structural Engineers	Mechanical Engineers	Electrical Engineers
16 ±	8 ±	4	2	2

With adequate supporting administration, these 32 built up to a total of 50–55 in the group as a whole. Experience showed this to be a good unit in which every member was known to the others and frequent contact, which made for good morale, was relatively easy. At the same time it became clear that the engineers were too few to generate adequate exchange of ideas and experience with their colleagues, while problems arose in their adequate representation at partner level. The ideal of each multi-disciplinary group having at least one partner from each of the main professions engaged would demand, for such a unit, a minimum of five partners. On the basis of a group of 32 professionals, this would represent a partner ratio of 1 : 5·5—well above the average of 1 : 8.* In other words the unit of 32 was not large enough to support an adequate partner structure, either in overall numbers or in professional representation. There would also be the further disparity as between one architect partner 'representing' sixteen architects in the group, and one mechanical or electrical engineer partner 'representing' two such engineers only.

Six factors, it was felt, had to be reconciled:

1. Adequate representation in the partnership
2. A partnership in balance with the firm as a whole
3. An adequate minimum number of members in any one profession to maintain interest
4. An overall size of group that would permit the maintenance of good personal contacts
5. A work load adequate to maintain income
6. A balance of work load to keep all professions busy

It was concluded that the best chances of reconciling these factors lay in a unit or group of 60–70 professionals, 110–120 overall, which would be capable of undertaking £2m.–£3m. of work each year and earning £250,000–£350,000 of fees.

The professional structure aimed at would be as follows:

<table>
<tr><th></th><th>Architects</th><th>Quantity Surveyors</th><th>Structural Engineers</th><th>Mechanical Engineers</th><th>Electrical Engineers</th></tr>
<tr><td>Partner</td><td>3</td><td>2</td><td>1</td><td>1</td><td>1</td></tr>
<tr><td>Associate</td><td>8</td><td>3</td><td>2</td><td colspan="2">2</td></tr>
<tr><td>Job Designer</td><td>21</td><td>11</td><td>5</td><td>2</td><td>2</td></tr>
<tr><td></td><td>32</td><td>16</td><td>8</td><td>4</td><td>4</td></tr>
</table>

* Figure from an RIBA Survey.

In working towards these larger groups, Building Design Partnership found it necessary, not least by reason of the illegality of more than twenty members in any one partnership, to reconsider their whole partnership structure. It was decided that there should be General Partners, figuring in all the groups, and Group Partners, attached to their group alone. As the former disengaged or retired, it was foreseen that they would normally be replaced from the ranks of the latter, though the possibility of bringing into the central, or General, partnership an outstanding outsider was not discounted. When, in 1964, the then seven offices were formed into three groups, the five existing partners became General Partners with overall responsibility for the firm as a whole; ten Associates became Group Partners, being jointly responsible for the offices comprising that group. The following year, after the opening of the Belfast office, the number of Group Partners was brought up to thirteen—four in Preston, three in Manchester, four in London, two in Belfast, with, in 1966, a further Group Partner in London and an Accountant-Secretary as a sixth General Partner.

The work of each office is co-ordinated by its Associates. The firm's philosophy allows for any number of Associates in an office: an Associate being defined as a professionally qualified person with the experience to act fully as an architect, surveyor or engineer would act in private practice. Thus in the London office of 70 (as distinct from the London Group, comprising London and Aldershot) composed of 25 architects (21 qualified), 8 quantity surveyors, 10 civil and structural, engineers, 11 mechanical and electrical engineers, 3 technicians, 1 librarian, 8 secretaries, 1 clerk, 1 comptometer operator, 1 telephonist and 1 general duties man, there were in 1966 four Group Partners (two of them architects, one a quantity surveyor and one a mechanical and electrical engineer) and eight Associates (five architects, one quantity surveyor and two engineers). Each office works in as self-sufficient a way as possible. Apart from accounts and some bulk buying, as much of the administration as possible is given to all the offices; for instance central communications is based on London.

Of particular interest is the unique salary system which Building Design Partnership has evolved. Briefly, a fixed number of points, 'mechanical shares', are given to each member for his various qualifications and office experience; a variable number of points, 'partnership shares', are additionally awarded according to each person's ability, responsibility, potential, poise, profitability and so on. The value of the share is determined, annually, by dividing the firm's income by the total number of shares allotted. Thus the value of the share was £30 in 1959–60, £33 in 1961–2, £35 in 1963–4.

The advantages of the system are seen as these. Everyone's salary is a multiple of the share so that everyone can contribute to the profitability of the firm, everyone can participate in the reward. There is an automatic rise each year because each person will have logged up another year's experience; further shares may be gained by increased responsibility or improvement; there is the bonus of the possibility of increase in the value of a share. Everyone knows that their salary is based on a system—which is used as a broad basis rather than rigidly fixed—which fairly relates everyone's salary whatever his profession.

Leadership and Rewards: II

by Misha Black

Group practice, in its essence, depends on those in the group *believing* that they are within that compass. They must believe that their own contribution to the group is an active factor in its existence and that for them to leave it would reduce its effectiveness. Their loyalty to the group must be such that they are as much concerned with its well-being and reputation as with their own personal glory. They must willingly acknowledge that the leadership within the group structure is fairly assumed, not imposed, and capable of being changed if it should deteriorate into ineffectualness. They must accept that their salaries properly reflect their individual contributions to the group and that, when profits are made, these are fairly distributed. In partial contradiction to these personal beliefs and attitudes, the group, if it is to have any sustainable validity, must be stronger than its parts, and thus able to continue and to grow even if many of those who are its constituent elements leave it. It must be capable of mutation and development irrespective of personalities.

Many architectural and design offices whose public *persona* is built around the names of their originators are in fact group offices innocently disguised. To come within the orbit of this book, however, they must exhibit one distinguishing feature: the creative originating work done by the group must be a distributed activity, the office must not be an association of a creative queen bee with draughtsmen executants diligently obeying the royal command. Shared creative responsibility is the basic requirement of group organization.

Creativity is not only the concern of those who practise the fine and the useful arts. The difficulty, however, of reconciling the problems of creativity with those of daily work and efficient organization are magnified in those professions which overtly and proudly announce creativity as their *raison d'être:* design, used in the sense which includes architecture,* clearly falls into this category. That some aspects of design have proved to be capable of objective analytical solution does not, at least for the present, diminish the argument. Even if it should, during the next decade, be shown that the scope

* I have used 'architecture' and 'design' as though they were interchangeable terms because, in their relation to group practice, the qualities they share are more important than those in which they differ. The practice of architecture and industrial design in all their varied categories is basically sufficiently similar to allow me to use the term 'design' comprehensively.

for creative decision is more constrained than most designers today accept, the fact remains that the contribution which every designer wants to make, which alone will satisfy his egoism, is a creative one, lying in his capacity for making new hypotheses from the base of fact, knowledge and experience. Every hypothesis must be new, for no moment in history can exactly repeat itself. The designer himself is being modified by time as he applies himself to a new task; his contribution to society depends on his capacity to match his new self to a new problem and, taking a creative step forward, to produce a solution which is ahead of the immediate historical moment.

That creativity is the foundation of their work is the faith which motivates all designers. Each will differently evaluate his creative ability, objective truth not always being reflected in subjective self-appreciation, but, to exist at all as a designer, a belief in at least a modicum of personally possessed creative ability is as necessary as it is for a brick-layer to have confidence in his capability of laying a straight course of bricks. The group structure, if it is to survive, must therefore allow each of its members to make their individual creative contribution: that is easily done in the early days of a group's life, it becomes more difficult as its adolescence progresses towards maturity. Early death is the inevitable fate of groups which do not quickly discover how effectively to distribute creative responsibility.

Design groups are originated in two primary forms: they are assembled by young designers who share the same attitudes and who believe that co-operation is psychologically, socially and organizationally more profitable than personal isolation, or they are groups of designers brought together by external events to achieve a known goal within a relatively short period. The first type of group is open-ended, believing in a limitless future; the second is closed, knowing that its dissolution is predetermined.

Creative responsibility is no immediate problem to the group of young men and women self-launched into their professional careers. They would not have joined together if they did not share compatible opinions about design form. Their arguments are about detail and practicability and not about principles. Being a small group they normally work at high pressure, they are their own draughtsmen and tea-makers, they are ennobled by youth and conviction and protected by innocence. Such groups can function as so completely integrated organisms as to produce work which retains the unity and individuality that characterizes a major work of art. But no organism can live without growth and group organizations are as susceptible to the laws of ageing and death as are all other biological systems.

It soon becomes apparent, firstly, that creativity is not equally shared between members of the group; secondly that efficient business organization is essential to the group's survival; thirdly that people working together for any length of time rub uneasily against each other; and lastly that those who believe they make the larger contribution to the well-being of the group will expect comparable recognition and financial gain. If the formation of the group is the thesis, its initial success is its antithesis, and a synthesis must be found if early dissolution is to be avoided. It is, in fact, exceptional for groups to survive for more than a few years; when they do it is because their members understand that controlled friction is the basis for group organization. As Ludwig von Bertalanffy said in a wider context: 'What is essential is interaction. In a system the forces balance each other and therefore their co-existence leads to a relatively stable configuration that resists destruction.'

The emergence of differing degrees of creativity is a wedge which can, and usually does, shatter the group structure. It creates a loaded introspective condition which is exacerbated by few, if any, members of the group being prepared to admit, even to themselves, that their imagination and inventiveness are less potent than those of others of their colleagues with whom they started to work on an assumed common plateau of ability.

During that period of re-adjustment, group survival can be achieved only by the patient human understanding of all who are involved. It can best be accomplished by dividing the previous mono-cellular structure of the group into single cells, each remaining in orbital contact but each separately responsible for individual jobs. The results eventually create their own moments of truth in which each sees his capability reflected in the grimly clear mirror of the completed commission. There is no room for self-deception in the ultimate confrontation between the designer and the results of his conception. The designer who is worthy of that designation sees the flaws in his completed work with greater clarity than do even his least sympathetic critics. The process may need to be repeated several times until the dialogue between pretension and reality is completed, but at its end each member of the group will have passed through the agony of adolescent creativity, will accept a rank order of ability and be prepared to acknowledge leadership. Those who do not then accept the general consensus will leave the group convinced of its unfairness.

From that stage onwards leadership is the key to the group's development. There is no alternative to this; those groups which claim that they work as an undirected co-operative mean only that the leadership is not openly acknowledged, that the leader is prepared to operate in anonymous disguise. But group leadership cannot be imposed—if it is, then the group no longer exists; it will have changed into another directed design office.

At the beginning of the first stage of the group's maturity the leader must be respected for his creative ability, nothing else will do. He will make the final decisions and they must be willingly accepted as being of the same order (that is based on creative understanding and conviction) as those which motivate all members of the group.

It is then, when the group has jettisoned immaculate ideas of equality, when it has thrown up its leader, when a hierarchy of creative talent has been recognized, that the group will be at the height of its power. It will have lost the capacity for producing the completely integrated statement which was the prerogative of its youth, but it will be able to design with that mature authority which serves well all conditions which do not require the nobility of individual genius.

There is a beguiling argument that the leader of a design group should be democratically elected by those whom he will later direct, and many attempts have been made to implement that idealistic concept. Every permutation of human relationship remains a theoretical possibility but I have never seen a system of elected leadership work satisfactorily in design practice for more than short periods. The group usually originally forms itself around the nucleus of one person or around a few who have been drawn together by shared affinities. Thereafter the initial core remains the generating force and its centrifugal momentum is stopped if attempts are made to change the organizational pattern to one sufficiently static and consolidated to allow for the holding of elections—and able to absorb the intrigue and jockeying for position which such elections inevitably imply. That I do not favour traditional democracy in the special situation of creative design groups does not lessen my support for political democracy; in that wider field the checks and counter-checks which democratic procedures entail are essential restraints on social irresponsibility and personal vainglory. In the small design group, however, democratic controls would quickly destroy the creativity which alone justifies its continuance.

But if the leader cannot be deposed by vote he remains exposed to a final irrevocable control: the members of the group can resign from it and leave the leader without a company to lead. This final sanction, and the break-up of the group which it implies, is only avoided when the leader is so engaged in the corporate life of the group, and has so identified his desires and ambitions with those of the group, as to make the growth of the corporate body more important to him than individual success. Such leadership will not be an authoritarian bending of the group to its ambition, but rather a sensitive interpretation of the needs of the group, able to guide it towards its known or only indistinctly perceived goal. If leadership is accepted, then the progressive devolution of leadership becomes possible as the group increases in size.

The need remains for the group to be able to change its leadership when this becomes calcified and no longer able to adapt itself to a changed situation. In the ideal situation the group leader will himself appreciate his failing capacity for creative control and move aside to allow others to assume his autocracy. The erstwhile leader may then become a wise counsellor whose experience and judgement will predict and disclose the traps of impracticability which threaten the less experienced and more creatively exuberant. If the leader insists on retaining creative leadership when he no longer has the capacity to do so, then the group will dissolve or become yet another humdrum design organization with a reputation for experienced practicability as partial compensation for its lack of creativity.

If a hierarchy of leadership must exist then its special responsibility must be recognized by the financial rewards being equally hierarchical. There is no alternative to this in our present society, and I doubt the longevity of any social system which does not permit responsibility which is assumed and efficiently discharged to be directly rewarded. But if the members of the group are sensitive to the fair distribution of earnings, they are even more uneasy about the proper appropriation of credit for creative authorship. Designers, almost without exception, are hypersensitive to public approbation. The public whose approval they seek may be limited to the editors of architectural and design magazines or to an even more exclusive coterie but, if they believe that their creative contribution has been wrongly attributed to another, then their despair and fury is equal to that of a mother robbed of a loved child.

Those designers who assume a nonchalance about credits are only more experienced in disguising their emotions or already so famous as to be able to afford occasional magnanimity.

This need for public acknowledgement can be satisfied by rigid group discipline which forbids all personal credit, extinguishes the conscious wish for individual acclaim by the strength of the taboo and provides compensation in the sense of glory in all the group's works, irrespective of their creative origin. Alternatively the need for public acceptance can be provided by individual credit being given to all members of the group who participated in any project which later justifies publication.

Personal anonymity has the major advantage of allowing all credit to accrue to the group so that its reputation grows undiminished by changes in its personnel or internal structure, but I prefer the system of individual acknowledgement under the group umbrella.

There is an uncontrollable, if irrational, tendency for those who are outside the group to identify its creative achievements with single individuals irrespective of their real creative contributions. Published personal credits restrict that tendency and simultaneously erect criteria for individual conduct and responsibility within the group itself. In actual working practice the method of apportioning credit may go through tidal phases, from anonymity at the formation of the group, to individual acknowledgement and then a return to anonymity when the reputation of the group is so high that this seems more important to its members than any personal approbation. Idealism is an essential element in group activity, but that idealism is related only to the quality of the work produced, to the creative function of the group, to its shared concept of form and its relation to the needs of our society: such creative group idealism does not resolve the anxious cares of private life. Idealism is the essential bone structure of the group, but sentimentality is a disease which will quickly debilitate and finally destroy it. The working pattern of the group must be tough and fair. Wages must be adjusted to the value of work done, bonuses distributed with reasonable relation to the profits earned, slackers dismissed, financial reserves accumulated, business affairs efficiently conducted. No group achieves perfection in its business and administrative organization, but the closer it approaches that goal, the longer will be its productive life. Those in the group who undertake the less glamorous functions of business and financial organization are as essential to its life as are lungs and kidneys to the brain.

The growth of the group brings with it problems which are even more critical than those of efficient management. They arise from the necessity for integrating a relatively large number of technical and junior design staff into the group structure. The new recruits have little knowledge of the original passionate motivation of the group, no loyalty to its antecedents and are not unreasonably initially suspicious of group talk in a situation which seems to them not very different from any conventional architectural or design office.

If the organization of the group has, however, kept pace with its growth without forsaking its original principles, the young designers soon become aware of its basic concept of shared opportunity for design inception and the shared public recognition which is its concomitant. It may then be wearisome for the older members of the group to deal with the unannealed enthusiasm of its new younger members, but only if they are allowed active creative participation will they develop into designers who will bring new strength to the group as a whole. Those who direct the group must, at this stage in its development, be constantly aware of the need to retain the essential elements of the group concept while the pressures of time schedules and expediency concert to turn it into a master-and-servant establishment.

Minor matters of nomenclature become important, the opportunity for the open discussion of design problems is essential, an atmosphere in which the most junior designer feels free to challenge the assumptions of his elders is necessary, opportunities for human contact are required, an atmosphere must be produced in which all feel that they are participating creatively within the compass of the group. A climate of shared endeavour and participating is not easy to engender amidst the pressures of a successful practice, but it can be done at least to a degree which will persuade some who join the group in its maturity to stay with it: others will inevitably find the atmosphere uncongenial and leave.

In the end the group will outlive its usefulness and should dissolve. There is comfort in long-established practices, and convenience in the sturdy administrative structures which build up round them, but the function of design is to find new formal relationships which simultaneously serve the needs of society and symbolize the emotional forces which motivate it. When the design group is no longer expressive and becomes content to reiterate forms which have only archaic interest, then its life is ended and its continuance is of value only to those who may make a livelihood from its efficiently serving their less adventurous clients.

Complacent middle age with its desire only to regurgitate the victories of its long-past youth can afflict groups as easily as it can devastate individuals. The group has greater resistance to atrophy by its capacity to rejuvenate itself as the younger designers take over creative responsibility from the older, but however well organized the creative leap-frogging may be, the time will come when the whole mechanism of the group is too well oiled, when its accumulated successes have established precedents for every new problem. That is the moment to call for the coffins in which quietly to lay the bones of a group which, having been drained of its vitality, is ready for honourable burial.

A danger in appreciating and predicting this ageing process lies, however, in attempts to counteract it by a desperate seeking for superficial originality when the position is not, in fact, that the group has aged but that it has moved into a new historical period which it does not properly comprehend. There are moments in the history of design when a new social situation demands expression by forms and the associations of forms as novel as the emergent situation itself; there are other times when the refinement of existing forms and the more subtle relating of them are the only valid creative acts.

In the quieter situation maturity may be mistaken for atrophy and epileptic jerks assumed to indicate vitality. The group must be able to adapt itself to the historical process of which it is a microcosm, to decide when it is useful to explore uncharted paths for the excitement of the excursion and when it should remain on the wider, if duller, road of additive progression. It must appreciate, as Sir Herbert Read has said, that 'it is peculiar to original genius to become less and less striking in proportion to its success in improving the taste and judgement of its contemporaries'.

The group must be constantly critical of its creativity, able to distinguish between the living and the dead, and self-analytical enough to be able to decide when new life can and must be injected into it, and when there is no alternative to old age and death being serenely accepted.

This problem of adjustment to ageing does not exist when groups are established for short-term projects: for the design of a group of buildings, or of an interior complex, or for the planning of an exhibition. It is in such undertakings that the maximum group creativity can rightly be expected. Members of the group are unconcerned by problems of a joint future which does not lie within the group framework: their energies are addressed to a known limited end.

In these circumstances the importance of leadership is magnified, the problems of personal recognition and private gain exacerbated and personal frictions exaggerated. The group may need to work at great and sustained pressure, everyone is conscious that the need is only for the group to hold together for a year, or for two or three years, after which the strains set up within its corporate body may be allowed to fragment it. Working in this situation is stimulating, exhausting and completely satisfying—if more so in retrospect than during the emotional creative tempest itself.

These short-term groups may be multiplied until they become a number of different groups each spinning vigorously on its own axis and yet within a single solar system with a co-ordinating centrifugal centre. This has been the pattern for some town planning projects and has reached a strange ephemeral maturity at major exhibitions where creativity and originality are momentarily more important than solid worth. This pattern is repeated on a smaller scale when a complex of adjoining buildings needs to be conceived simultaneously by different architects, or when several groups of designers are working at the same time on related engineering products or on the design of different rooms in the same building.

In such solar systems of group activity the need for unity in diversity overpowers all other considerations. The individual groups themselves must have a bond of sympathy and understanding with each other—if they lack this the project will fail, to the degree that any of the groups may be outside the continuum of shared beliefs and attitudes. But, given a generally shared approach to design, the individuality of each group will ensure the organic visual diversity essential to the success of any large-scale project.

In such conditions the task of the central co-ordinator, once the master-plan has been devised, is only to oil the bearings, cool the overheated creative engines when their tops threaten to fly off, and ensure that the perimeter of one group's area meshes easily with the zone where another group's responsibility begins.

Diverse and antagonistic personalities can be assembled even within a group whose aim is to live for longer than the time needed to complete a single assignment. If a permanent group numbers more than some ten people, the division of the group into cells which can be formed, dissolved and re-formed in different permutations is essential. Only thus will it retain its vitality. Antagonism and competition between the cells is to be encouraged, but the tensions must be kept below surface so that the group can continue to function at the overt level of business efficiency and tolerable human relations. Design groups should not be happy families acquiescent to a father figure leader. They should be disturbed by theoretical differences and opposing creative convictions, their members should be passionate in their criticism of their colleagues' work and hot tempered about their personal rights and dignity. The basis of group organization is that it is a voluntary assembly of people: no one is compelled to stay in the group, anyone may leave it at any time. The fact that a group stays in existence is proof of its validity, but its calm outer face must conceal inner conflicts. If its creative working interior also is an undisturbed calm, then it has become as lifeless as a dead sputnik, still orbiting aimlessly round the world.

References

For the most part, titles listed are related to specific references in the text. Here and there an additional item has been allowed to creep in because its more general application to the subject appears of particular value. The resulting list makes no claim to be a complete or balanced bibliography.

CHAPTER 1

The Social History of Art. Arnold Hauser. Two vols. Routledge & Kegan Paul, London, 1951 (paperback edition 1963). Many of the matters touched on in this chapter are developed in detail by Dr Hauser in this exhaustive and authoritative work.

The Classical Language of Architecture. Sir John Summerson. BBC, London, 1963. Reprinted.

'The Chronicle of Monte Cassino'. Leo of Ostia (1046?–1115), librarian at Monte Cassino and later Cardinal Bishop of Ostia. Translated by Dr Herbert Bloch and included in *Literary Sources of Art History,* Princeton University Press, New Jersey, 1947. One of the most reliable accounts of building work in the Middle Ages. Other informative accounts include Abbot Suger (1081–1151) on the affairs of the Abbey of St Denis and the account of the burning and repair of the Church of Canterbury by Gervase of Canterbury (1141–1210) in his *Chronica* covering the years 1100–1199.

De re aedificatoria (*On Architecture*). Alberti, 1485. Translated by Giacomo Leoni (from an Italian translation of the Latin text) as *The Architecture of Leon Battista Alberti.* London, 1726.

The London Tradesman. Campbell, London, 1747.

Architecture in Britain, 1530–1830. John Summerson. Penguin Books, Harmondsworth, 1953.

Georgian London. John Summerson. Pleiades Books, London, 1945. Revised edition Pelican Books, Harmondsworth, 1962.

Town Planning in London. Donald J. Olsen. Yale University Press, New Haven, 1964. An account of how the great London estates were developed during the eighteenth and nineteenth centuries.

CHAPTER 2

Victorian Architecture. Robert Furneaux Jordan, Penguin Books, Harmondsworth, 1966. An exceptionally readable general survey of the period.

Encyclopaedia of Cottage, Farm and Villa Architecture. J. C. Loudon, London, 1833.

True Principles of Christian Architecture. A. W. N. Pugin, London, 1841.

Apology. A. W. N. Pugin. London, 1843.

'Pre-Raphaelite Painted Furniture'. Michael Archer. Article in *Country Life,* 1 April 1965.

William Morris. Selected Writings and Designs. Edited by Asa Briggs. Pelican Books, Harmondsworth, 1962. A useful introduction to the range of Morris's ideas.

'Gropius and Van de Velde'. Nikolaus Pevsner. Article in *The Architectural Review,* 793, March 1963.

Pioneers of the Modern Movement. Nikolaus Pevsner. Faber and Faber, London, 1936. Revised for Penguin Books, Harmondsworth, 1960, as *Pioneers of Modern Design.* The standard work on the subject and essential reading.

REFERENCES

Theory and Design in the First Machine Age. Reyner Banham. Architectural Press, London, 1960. A detailed sequel to Pevsner's *Pioneers*.

De Stijl. Vol. I, 1918. Vol. III, 1920. Vol. IV, 1921.

Bauhaus 1919–1928. Gropius and Bayer. New York, 1938. For English texts of Bauhaus documents.

Hannes Meyer. Claude Schnaidt, Niggli, Teufen and Tiranti, London, 1965.

'Open Letter to an American Student'. Colin St. J. Wilson, in *Program No. 3*. School of Architecture, Columbia University, NY.

Royal College of Art Lethaby Lecture 1965. Dr Tomas Maldonado. Delivered RIBA, March 1965.

Ulm 8/9, Hochschule für Gestaltung, Ulm-Donau, 1963.

Vers une Architecture. Le Corbusier. Paris, 1923. Translated by Frederick Etchells as *Towards a New Architecture*, 1927, reprinted by the Architectural Press, London, 1946.

CHAPTER 3

Art and Industry. Herbert Read. Faber, London, 1966.

The Nature of Design. David Pye. Studio Vista, London, 1964.

Report of Conference on Design Methods. Edited J. C. Jones and D. G. Thornley. Pergamon Press, London, 1963.

Design Management. Michael Farr. Hodder & Stoughton, London, 1966.

The Measure of Man. Human Factors in Design. Henry Dreyfuss. Witney Library of Design, Hill and Wang, New York, 1959.

Human Factors Engineering. Ernest J. McCormick, McGraw-Hill, New York, 1965. A revised and expanded edition of the same author's *Human Engineering* of 1957.

Drawing for Illustration. Lynton Lamb. Oxford University Press, London, 1962.

'The Illustrations for *The Forsyte Saga*'. Anthony Gross. *The Penrose Annual*, Lund Humphries, London, 1951.

'The Role of the Architect-Planner and the Architect in Planning'. Address by Sir William Holford to the Architectural Association, 4 December 1962. AA *Journal*, January 1963.

Systematic Method for Designers. Bruce Archer. A series of articles in *Design*, commencing April 1963, and subsequently reprinted in book form. Council of Industrial Design, London, 1966.

'Design Method in Architecture'. Geoffrey H. Broadbent, *Architects' Journal* 14 September 1966. This article led to further discussion in the issue dated 21 December 1966.

Notes on the Synthesis of Form. C. Alexander, Harvard University Press, Cambridge, Massachusetts, 1964.

'Paper prepared for Management Advisory Group'. John Madge. RIBA *Journal*, November 1964.

CHAPTER 4

Survey of Problems Before the Construction Industry. Report by Sir Harold Emmerson, Her Majesty's Stationery Office, London, 1962.

Report of the NJCC *Committee* (the 'Noel Hall Committee') *on Training in the Building Industry*. RIBA, London, 1964.

'Building Research—Current Papers'. Design Series No. 6—Reprinted from RIBA *Journal*, 1963.

'Comprehensive Architectural Practice for Small Offices or Large'. Dudley Hunt Jnr. AIA *Journal*, Washington, September 1962.

Design, 183, March 1964. A special issue on America, edited by Ralph Caplan.

The Psychology of Study. Professor C. A. Mace. Revised edition, Pelican Books, Harmondsworth, 1962.

The Architect and his Office. A survey published by the Royal Institute of British Architects, London, 1962.
Guide to Group Practice and Consortia. Royal Institute of British Architects, London, 1965.
'The Architects from "Skid's Row"'. *Fortune,* January 1958.
'The Architects' Collaborative'. Sarah Harkness. *Architectural Review,* May 1957. (See also Gropius, AR, September 1953.)
The Practice of Management. P. F. Drucker. Heinemann, London, 1961.

CHAPTER 5
'The Festival of Britain 1951'. Three Cantor Lectures by Sir Gerald Barry at the Royal Society of Arts, May 1952. RSA *Journal,* August 1952.

CHAPTER 6
Designing for People. Henry Dreyfuss. Simon and Schuster, New York, 1955.
'Industrial Design a Vital Ingredient'. William F. H. Purcell, *Automotive Industries,* 15 May 1961. An account of the design work on the John Deere range of tractors.
'Men and Machines on the Farm'. Professor W. F. Floyd, *Design,* April 1967.
Case histories appear regularly in *Design* magazine.

CHAPTER 7
'LCC Architect's Department'. Article by Gontran Goulden in *The Architect and Building News,* 31 March 1965. A brief 'obituary' on the demise of the Department.
Detailed job histories appear from time to time in the *Architects' Journal.*

CHAPTER 8
'Islamabad, the New Capital of Pakistan'. Doxiadis Associates, *Ekistics,* November 1964.
'Islamabad. The creation of a new capital'. C. A. Doxiadis, *The Town Planning Review,* University of Liverpool, April 1965.
'The Design of Islamabad'. The Sir George Birdwood Memorial Lecture, by Derek Lovejoy at the Royal Society of Arts, 17 May 1966, RSA *Journal,* October 1966.
Islamabad. New Dimensions in Architecture. Khawaja Zahir-ud-Deen. Capital Development Authority, Rawalpindi, July 1964.

CHAPTER 9
'A Lovely Ship . . .'. By Charles Morris. *The Vickers Magazine,* Summer 1957.
'The Interior Design of Passenger Ships'. Thomas Gray Memorial Lecture by Sir Colin Anderson at the Royal Society of Arts. RSA *Journal,* May 1966.

CHAPTER 10
Television by Design. Richard Levin. The Bodley Head, London, 1961. A detailed study of the problems of designing for television.

CHAPTER 11
'Housing Experiment at Peterlee'. Symposium and discussion at the Architectural Association, 22 February 1961. AA *Journal,* June 1961.
Communications in the Building Industry. Dr Gurth Higgin and Dr Neil Jessup. Tavistock Publications, London, 1965.
'The Architect in Society'. An address by Walter Gropius, 21 March 1961. Reprinted in *Program.* School of Architecture, Columbia University, N.Y., Spring 1961.
Bulletin 41 of the International Association of Plastic Arts. UNESCO,
The Human Group. G. C. Homans. Routledge and Kegan Paul, London, 1951.
Individual in Society. Krech *et al.* McGraw-Hill, New York, 1962.
Human Groups. W. J. H. Sprott. Penguin Books, London, 1958.

Acknowledgements

Thanks for providing illustrations are due to those listed below. References are to page numbers; position on page is indicated as follows:

t, c, f = top, centre, foot
l, r = left, right
p. = photographer

17 t, A. F. Kersting
17 f, from *The Pattern of English Building* by Alec Clifton-Taylor, B. T. Batsford Ltd.
18 t, 19, 28 tl, 28 fl, J. Allan Cash
18 f, Central Office of Information
20 l, 21 f, from *Italian Villas and Palaces* by Georgina Masson, Thames and Hudson Ltd
20 r, Mansell Collection
21 t, from *Schönbornschlosser* by Karl Lohmeyer (Widener Library, Harvard University)
22 t, Courtauld Institute of Art
22 f, Monotype Corporation
23, German Tourist Information Bureau
24, 47, Victoria and Albert Museum
25, p. A. C. Cooper, Ltd. (owner unknown)
27, 77 tr, tl, Science Museum
28 tr, 46 fr, 51 tr, 56 t, 115 f, Architectural Press, p. H. de Burgh Galwey
28 fr, Victor Rose
45 tr, Architectural Press, p. Kenneth Browne
45 fl, Architectural Press, p. Stewart Bale
45 fr, National Monuments Record
46 t, Brian Reade
46 fl, Architectural Press
48, Stedelijk Museum, Amsterdam
49, Gemeente Museum, the Hague
50 tl, 51 tc, 185–8, Studio Ponti-Fornaroli-Rosselli
50 tc, David Mellor
50 tr, Kenneth Grange
50 f, Radio Times-Hulton Picture Library
51 tl, Council of Industrial Design
51 f, Ekholm and White
52, Architectural Press, p. Cervin Robinson
53, 59 tl, Cement and Concrete Association, p. Sidney W. Newbery
54, Institute of Contemporary Arts
55 t, Alexander Girard, p. Charles Eames
55 f, 150 t, 151 fr, 152, 153 f, 155, 156, 157, 159 f, 160, Greater London Council
56 f, © Ezra Stoller (ESTO)
57, from *Henry Moore, Sculpture and Drawings*, vol. iii, edited by Alan Bowness, Lund Humphries
58 t, Museum of Fine Arts, Houston
58 f, Denys Lasdun and Partners, p. John Donat

59 tr, Maidenhead Brick and Tile Co., p. Jane Gate
59 cr, Max Braun Co.
59 fl, Architectural Press, p. Eric de Maré
59 fr, Cabinetmaker Johannes Hansen, architect Hans J. Wegner
60, Peter Womersley, p. Waverley Studios
77 f, University of Reading, Museum of English Rural Life
78, 125–32, Henry Dreyfuss
79 t, Royal Horticultural Society
79 fl, r, from *The Mini Story* by Lawrence Pomeroy, George Newnes Ltd. (The Temple Press)
80 t, f, Société Mondiale Villevorde (Council of Industrial Design)
81 t, f, 82 t, c, British Rail
82 fr, Civic Trust
83 t, Ezra Stoller
83 f, Victor Pasmore, p. John Pasmore
84 t, Arup Associates, p. Colin Westwood
84 f, Ministry of Public Building and Works, p. Crispin Eurich
109 t, f, 110 t, 111 f, Misha Black
110 f, 111 t, Peter Kneebone
112, 113, 114, 115 t, 116 t, Henrion Design Associates
116 f, Aerofilms
149, 150 f, 151 tr, tl, fl, 153 t, 158, 159 t, *Architect and Building News*, p. Henk Snoek
154, Greater London Council, p. R. W. Robson-Smith
205, 206 t, Vickers-Armstrongs
206 f, 211 f, 212, P & O Orient Lines
207–11 t, Design Research Unit, p. John Maltby
237–48, reproduced by permission of the British Broadcasting Corporation

Index

As already stated, this book attempts no more than to consider the design process in a broad context. The index, accordingly, is not exhaustive but provides signposts only to the more important themes. Italic numbers signify illustrations.